Modern Control

State-Space Analysis and Design Methods

Arie Nakhmani, Ph.D.
University of Alabama at Birmingham

New York Chicago San Francisco
Athens London Madrid
Mexico City Milan New Delhi
Singapore Sydney Toronto

Library of Congress Control Number: 2020934135

McGraw Hill books are available at special quantity discounts to use as premiums and sales promotions or for use in corporate training programs. To contact a representative, please visit the Contact Us page at www.mhprofessional.com.

Modern Control: State-Space Analysis and Design Methods

Copyright © 2020 by McGraw Hill. All rights reserved. Printed in the United States of America. Except as permitted under the United States Copyright Act of 1976, no part of this publication may be reproduced or distributed in any form or by any means, or stored in a data base or retrieval system, without the prior written permission of the publisher.

1 2 3 4 5 6 7 8 9   LCR   25 24 23 22 21 20

ISBN 978-1-260-45924-1
MHID 1-260-45924-1

This book is printed on acid-free paper.

Sponsoring Editor
Lara Zoble

Editorial Supervisor
Stephen M. Smith

Production Supervisor
Lynn M. Messina

Acquisitions Coordinator
Elizabeth M. Houde

Project Manager
Tania Andrabi,
Cenveo® Publisher Services

Copy Editor
Surendra Shivam,
Cenveo Publisher Services

Proofreader
Alekha C. Jena

Indexer
Arc Indexing, Inc.

Art Director, Cover
Jeff Weeks

Composition
Cenveo Publisher Services

Information contained in this work has been obtained by McGraw Hill from sources believed to be reliable. However, neither McGraw Hill nor its authors guarantee the accuracy or completeness of any information published herein, and neither McGraw Hill nor its authors shall be responsible for any errors, omissions, or damages arising out of use of this information. This work is published with the understanding that McGraw Hill and its authors are supplying information but are not attempting to render engineering or other professional services. If such services are required, the assistance of an appropriate professional should be sought.

To my parents, who have always been there.

About the Author

Dr. Arie Nakhmani (M.Sc. in Robust Control, Ph.D. in Computer Vision) is Associate Professor of Electrical and Computer Engineering, Associate Scientist in the Comprehensive Cancer Center, and Director of the ANRY Lab at the University of Alabama at Birmingham. He is the author of over 50 peer-reviewed research articles and book chapters on robust control, machine learning, and signal and image analysis.

Contents

	Preface	ix
	Acknowledgments	xi
1	**Introduction to Control Systems**	**1**
	Control System Design Goals	1
	Plant Structure	2
	Modeling	3
	Conversion from ODE to Transfer Function	4
	Conversion from Transfer Function to ODE	6
	Stability	7
	State-Space System's Representation	10
	Why Should We Learn about State Space?	11
	Conversion from ODE to State Space	11
	Solved Problems	13
2	**State-Space Representations**	**17**
	Continuous-Time Single-Input Single-Output (SISO) State-Space Systems	17
	Transfer Function of Continuous-Time SISO State-Space System	18
	Discrete-Time SISO State-Space Systems	20
	Transfer Function of Discrete-Time SISO State-Space System	20
	Multiple-Input Multiple-Output (MIMO) State-Space Systems	21
	Stability of Continuous-Time Systems	21
	Stability of Discrete-Time Systems	21
	Block Diagrams of State-Space Systems	22
	Controllability	24
	Observability	25
	Minimal Systems	25
	State Similarity Transforms	25
	Canonical Forms	27
	Solved Problems	30
3	**Pole Placement via State Feedback**	**35**
	State Feedback	35
	Controller Design	37
	Tracking the Input Signal	38
	Integrator in the Loop	38
	Solved Problems	40

4 State Estimation (Observers) — 43
Observer Structure — 43
Observer Design — 45
Integrated System: State Feedback + Observer — 46
Solved Problems — 50

5 Nonminimal Canonical Forms — 57
Canonical Noncontrollable Form — 57
Canonical Nonobservable Form — 58
Stabilizability and Detectability — 59
How to Check Controllability and Observability of Eigenvalues — 59
Solved Problems — 60

6 Linearization — 69
Equilibrium Points — 69
Solved Problems — 71

7 Lyapunov Stability — 79
Internally Stable Systems — 79
Direct Lyapunov Method (Second Method) — 80
Lyapunov Stability for Continuous-Time LTI Systems — 81
Lyapunov Stability for Discrete-Time LTI Systems — 82
Solved Problems — 83

8 Linear Quadratic Regulators — 87
Cost Function — 87
Continuous-Time Optimal Controller — 88
Cross-Product Extension of Cost Function — 88
Prescribed Degree of Stability — 89
Discrete-Time Optimal Controller — 89
Solved Problems — 90

9 Symmetric Root Locus — 95
Continuous-Time SRL — 95
Discrete-Time SRL — 97
How to Sketch Continuous-Time SRL — 97
How to Sketch Discrete-Time SRL — 98
Solved Problems — 98

10 Kalman Filter — 105
The Idea of Optimal Observer (Estimator) in Presence of Noise — 105
Optimal Observer (Kalman Filter) — 108
Recursive Solution — 110
Alternative Kalman Filter Formulation for Unknown
    Initial Conditions — 111
Solved Problems — 112

11 Linear Quadratic Gaussian Control **121**
  Kalman-Bucy Filter 121
  What Is LQG Control? 122
  Optimal Cost Function for Stationary LQG 123
  Solved Problems 124

12 Project Examples **129**
  General Instructions 129
  Project 1: Magnetic Levitation System Control 129
  Project 2: Double Inverted Pendulum 133
  Project 3: Bridge Crane Control 140

A Math Compendium **147**
  A Notation and Nomenclature 147
  B Trigonometric Identities 147
  C Complex Numbers 148
  D Algebra 148
  E Calculus 150
  F Signals and Systems 153
  G Linear Algebra 160
  H Random Variables 170

References **173**

Index **175**

Preface

The aim of this book is to help with your journey into state-space methods of modern control. It could be said that modern control marked its beginning when the first satellite was launched into space by the former Soviet Union on October 4, 1957. The techniques that were used to control that satellite were very different from the classical control. They were based on state-space models in the time domain and on the stability theory developed by Aleksandr Lyapunov (Lyapunov, 1892). A state-space approach has multiple advantages over the classical transfer function approach and allows:

1. Uniform analysis and design techniques for single-input single-output (SISO) and multiple-input multiple-output (MIMO) systems.
2. Almost-uniform analysis and design for continuous-time and discrete-time systems.
3. Relocating closed-loop poles anywhere you want (for minimal systems) by a simple computation without trial and error.
4. Development of optimal and robust control techniques.
5. Extension to analysis and design of nonlinear systems.

Additionally, a state-space approach does not assume zero initial conditions (as in transfer functions).

The state-space approach is widely used in multiple applications beyond modern control, such as visual tracking (computer vision), biomedical signal and image analysis, forecasting stock prices (econometrics), or machine learning. There are alternative names for state-space models in various fields of study. For example, in forecasting time series they could be called latent process models and in machine learning it is more frequent to hear about hidden Markov models (HMMs) (Baum, 1966).

Is This Book for You?

This book is appropriate for a one-semester upper-level undergraduate or graduate course, but could also benefit professionals who want to learn more about state space. While other textbooks strive to give you a comprehensive guide to numerous theorems of modern control and their proofs, I have made a humble attempt to show the available techniques concisely and explain where they come from. While currently most computations are done using a computer, I believe that it is still beneficial to know how those computations are done by hand, for a better understanding. Therefore, I added plenty of solved problems with an increasing degree of complexity to this book. I have also provided tips and tricks for solving exercises and interjected self-evaluation questions with answers. I recommend that you think well and try answering the questions on your own before reading the answers.

This book assumes basic familiarity with the classical control systems terminology, linear algebra, ordinary differential equations, and Laplace and Z-transforms since most of these topics are covered in undergraduate courses on signals and systems. Additional information on this math background is provided in the Appendix. While the book could be read and tools applied without sufficient knowledge of the mathematical theories, this knowledge is very useful for understanding how those tools were developed and why they work.

What You Will Find in This Book

This book is organized around a state-space approach to the analysis and design of control systems. It is important to understand the language of control theory in order to understand modern control. Clear motivation for the need for modern control comes from understanding the problems and deficiencies associated with the classical control design approaches. Thus, in Chapter 1, we outline the terminology and essential basics of classical control theory and connect it to a state-space representation. Chapter 2 describes linear systems theory, which is fundamental for the understanding of modern control. This chapter has a lot of math and might be difficult to read. I recommend reading it slowly, trying to grasp the main definitions, and returning to it later when needed. In Chapter 3, you will learn how to apply modern control theory while stabilizing a system if all states are measured directly by sensors. Unfortunately, in real life not all states are accessible and Chapter 4 describes how to estimate states without measuring them directly. Chapter 5 discusses the design methodology for systems with a cancellation in the transfer function. Chapters 6 and 7 give a brief introduction to handling nonlinear systems and analyzing their stability using Lyapunov's theory. Chapters 8 through 11 provide an introduction to optimal control by describing the design of an optimal linear quadratic regulator, its interpretation in the root locus domain, the Kalman filter, and the linear quadratic Gaussian controller which allows emulating a serial or feedback controller using state-space tools. Finally, Chapter 12 describes practical simulation projects or labs that could reinforce learning of theoretical topics.

The comprehensive Appendix at the end of the book provides formulas ranging from basic math to calculus, differential equations, linear algebra, and signals and systems, and will help you recap the material, learn something new, and survive through mathematically heavy courses.

Additional materials and the MATLAB code are provided online at https://www.mhprofessional.com/Nakhmani.

Arie Nakhmani, Ph.D.

Acknowledgments

My own journey into the world of control systems started with my amazing and charismatic teacher, the late Professor Raphael Sivan. He instilled in me a lifelong passion for feedback mechanisms. I am indebted to my mentors, Drs. Michael Lichtsinder, Ezra Zeheb, and Allen Tannenbaum, for teaching, encouraging, and inspiring me to get a deeper understanding of control topics. I am thankful to Drs. Nahum Shimkin, Arie Feuer, and Yossef Steinberg from Technion—Israel Institute of Technology, who I assisted in teaching the modern control course. I learned a lot from them and in many ways this book is inspired by their way of teaching. Some examples in this book have also appeared in tests given at Technion and the University of Alabama at Birmingham. I am grateful to my colleagues and students for fruitful discussions and unexpected questions that helped improve the material in this book. I also would like to thank the editors and reviewers at McGraw Hill for their valuable tips and for making sense out of my initial text.

Last but not least, I would like to thank my family for providing emotional support and not letting me dive too deeply.

CHAPTER 1
Introduction to Control Systems

This chapter provides a brief refresher on the classical control theory and the different mathematical prerequisites which will help you to better understand modern control. It is important to learn control systems analysis and design since many important applications exist, ranging from consumer electronics to space research, but the most important part you should take from any control systems material is a special mindset, where you think about the system as a black box without knowing specific details. Control theory allows you to analyze the system despite model uncertainty, disturbances, and noises. More than that, you can make this system work as needed, even if you have an incomplete knowledge of the system's model. We will apply many mathematical tools in this book; nevertheless, you should always remember that the control theory is all about physical devices. Thus, if the mathematical result contradicts the physical abilities of the system, the mathematical result is wrong! As an example, let's assume that a good athlete can run at a speed of 25 kilometers per hour. How long will it take for that athlete to run 100 kilometers? Mathematically, the correct answer would obviously be 4 hours, but physically the correct answer probably would be that no athlete can run that far that fast. Similarly, if you correctly designed a controller that is supposed to produce 1 million volts as an input to the controlled system, it would not work in real life. This example also shows the importance of using simulations before implementations.

Control System Design Goals

The system we want to control is called a *plant* (see Figure 1.1). The input signal (or signals) of the plant is called *control effort* and denoted by $u(t)$, and the output is denoted by $y(t)$.

The main goal of control theory is to design a controller $C(s)$ so that the system output $y(t)$ will closely track the given reference signals $r(t)$ (see Figure 1.2), despite disturbances and noise entering the system.

Typically, the system should be stabilized about some design point, and the response to the disturbances $d(t)$ and noise $n(t)$ should be reduced. An additional desired property of a closed-loop system is to have control effort $u(t)$ as small as possible. Given the goals of control system design and all the desired properties discussed above, it seems trivial to design controllers, as will be explained in the following question. I recommend trying to answer all questions by yourself before reading the answers.

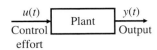

FIGURE 1.1 Schematic representation of a plant, control effort, and output.

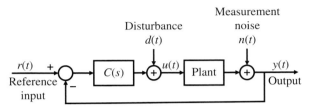

FIGURE 1.2 Typical closed-loop system with controller C(s).

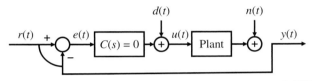

FIGURE 1.3 Allegedly ideal feedback system design.

Question on Control-Loop Structure

Okay, so we want the output signal to closely follow the reference signal. We can connect zero controller $C(s) = 0$ and then connect the reference directly to the output like in Figure 1.3.

This way we get zero error $e(t) = r(t) - y(t) = 0$ (by definition) and the output follows exactly the reference. Disturbances and noises do not affect the system and the control effort is zero. So, we have got an ideal control system, right?

Answer

The goal of control is not to make the system's output to follow the input, but to make a plant to produce the output which follows the reference input. The plant output is measured by sensors, and this is the real output of the system [and not the abstract $y(t)$ arrow to the right]. Since the plant is not producing output, this design is useless.

Plant Structure

A typical control system includes sensors and actuators (see Figure 1.4). *Actuators* are a type of transducers that convert input signal into physical energy that the plant can handle (e.g., motion, electricity, temperature). *Sensors* convert the measured output of a system into output signal (typically voltage).

It is important to remember that sensors and actuators might be coded in the computer or be an integral part of a plant. Selecting suitable sensors and actuators significantly simplifies the design of a control system.

FIGURE 1.4 Plant structure.

Modeling

The most popular way to model a system is by using the appropriate physical principles. Most systems have mechanical, electric, or chemical parts that can be modeled accordingly. There are other methods for system modeling. For example, if the system is stable in an open loop, we can try to measure step response of this system in the open loop. Based on the measurement of steady-state error, settling time, and overshoot, we can estimate the location of the dominant poles and the gain. Then we can tune other parameters by using the error minimization process (optimization). Another approach is to input sinusoidal waves of different frequencies and amplitude of 1 to the system (in the open loop), and to measure the output sinusoidal wave (for linear system, the output is always sinusoidal). The difference in the input and output phases and amplitudes constitutes the Bode diagram of the system. Using the Bode diagram, the transfer function can be estimated.

> **CAUTION!** For nonlinear systems, the step response as well as the Bode response is dependent on the amplitude of the input signal, and the superposition rule does not work anymore. Therefore, there is no sense in talking about the transfer function of a nonlinear system.

> **CAUTION!** Each measurement or parameter has a tolerance range (uncertainty); thus, the obtained (estimated) model is never precise.

Why Linear Systems?

Most (if not all) physical systems we know are nonlinear. We will have a few chapters discussing nonlinear systems, but the control theories we are learning in this book are mainly about linear systems. Tools like transfer functions or Bode diagrams are not appropriate for nonlinear systems, so why bother learning them?

Answer

It is much simpler to analyze and design linear systems. Many times, nonlinear systems could be approximated locally by linear systems.

Modeling Example

Let's develop a model for an angle control of a DC motor given in Figure 1.5 from physical principles.

The physics behind those equations is beyond the scope of this book, but what will be important for us is how to convert from the obtained differential equations into other representations.

4   Chapter One

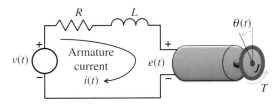

FIGURE 1.5   Direct current motor modeling.

The input of this system is the voltage $v(t)$, and the output is the measured angle $\theta(t)$. The torque T is proportional to the armature current:

$$T = Ki(t) \tag{1.1}$$

The back emf $e(t)$ is proportional to the angular velocity:

$$e(t) = K\dot{\theta}(t) \tag{1.2}$$

Note that in SI units, both coefficients of proportionality are equal (K).
Using Newton's second law, we obtain

$$J\ddot{\theta}(t) = T \tag{1.3}$$

where J is the moment of inertia of the rotor. By applying Kirchhoff's voltage law, we get

$$L\frac{di(t)}{dt} + Ri(t) + e(t) = v(t) \tag{1.4}$$

From (1.1) and (1.3)

$$J\ddot{\theta}(t) = Ki(t) \tag{1.5}$$

From (1.2) and (1.4)

$$L\frac{di(t)}{dt} + Ri(t) + K\dot{\theta}(t) = v(t) \tag{1.6}$$

The system of Equations (1.5) and (1.6) is our DC motor *ordinary differential equations* (ODEs) model (see Appendix F.11). To solve the equations, we need, of course, the initial conditions.

Conversion from ODE to Transfer Function

In classical control theory, for designing a controller we need our model to be represented as a transfer function. To convert ODE representation into the transfer function, three steps are necessary:

1. Apply Laplace transform to both sides of each ODE.
2. Solve the system of the obtained algebraic equations to get the output as a function of the input.
3. Assume *zero state response* (ZSR), that is, all the initial conditions are zero, and find the output to the input ratio [in Laplace (s) domain]. This ratio is the *transfer function*.

REMINDER Laplace transforms of first two derivatives of a function $f(t)$ are given by (Appendix F.10)

$$\mathcal{L}\{\dot{f}(t)\} = sF(s) - f(0)$$

$$\mathcal{L}\{\ddot{f}(t)\} = s^2F(s) - sf(0) - \dot{f}(0)$$

where $F(s)$ is the Laplace transform of $f(t)$.

Note that for computing transfer function, it is easier to assume ZSR first (in this case the Laplace transform is much simpler). In other words, we could apply the simplified Laplace transform $\mathcal{L}f^{(n)}(t) = s^n F(s)$ to both sides of ODE.

Example

The system's ODE is given by $\ddot{y}(t) + 5\dot{y}(t) - 10y(t) = 3\dot{u}(t) + 4u(t)$. To convert this representation into a transfer function, first we apply Laplace transform to both sides of the ODE by replacing each nth-order derivative with a multiplication by s^n and converting all variables to s domain:

$$s^2 Y(s) + 5sY(s) - 10Y(s) = 3sU(s) + 4U(s)$$

Now, we need a simple algebraic manipulation on this equation to find the transfer function which is defined by $\frac{Y(s)}{U(s)}$. By combining all multipliers of $Y(s)$ from the left and all multipliers of $U(s)$ from the right, we get the transfer function $\frac{Y(s)}{U(s)} = \frac{3s+4}{s^2+5s-10}$.

Example

In this example, we want to continue working with a DC motor model. This time we will not assume zero initial conditions to demonstrate a problem with the development of a transfer function from ODE. We apply Laplace transforms to Equations (1.5) and (1.6):

$$J(s^2 \Theta(s) - s\theta(0) - \dot{\theta}(0)) = KI(s) \tag{1.7}$$

$$L(sI(s) - i(0)) + RI(s) + K(s\Theta(s) - \theta(0)) = V(s) \tag{1.8}$$

After a few algebraic manipulations [solving (1.7) and (1.8) for $I(s)$ and $\Theta(s)$]

$$\Theta(s) = \frac{KV(s)}{JLs^3 + RJs^2 + K^2s} + \frac{K^2\theta(0) + J(Ls+R)(s\theta(0) + \dot{\theta}(0)) + KLi(0)}{JLs^3 + RJs^2 + K^2s} \tag{1.9}$$

The transfer function is defined as a ratio between the output $\Theta(s)$ and input $V(s)$. Since all initial conditions are represented by constants, the overall shape of Equation (1.9) is $\Theta(s) = \alpha(s)V(s) + \beta(s)$, where α and β are some expressions independent of $V(s)$ or $\Theta(s)$. It is clear from that shape that it is impossible to express $\Theta(s)/V(s)$ in terms of s only as needed for a transfer function. In other words, $\beta(s)$ must be identically zero to isolate the transfer function ratio $\alpha(s)$. This happens only for all initial conditions zero, or ZSR.

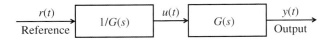

FIGURE 1.6 Allegedly perfect controller design methodology.

CAUTION! Transfer function is defined *only* for initial conditions zero; thus, it is not completely equivalent to the ODE representation. In a transfer function representation, we essentially ignore initial conditions (which in most cases we don't know and cannot control).

Now, for ZSR the rightmost part of Equation (1.9) is equal to zero; thus, the transfer function is

$$G(s) = \frac{\Theta(s)}{V(s)} = \frac{K}{JLs^3 + RJs^2 + K^2 s} \tag{1.10}$$

Question on Control-Loop Structure
We know that the open-loop systems are sensitive to disturbances, but let's forget that for a minute.

I propose to design the open-loop controller to the plant in Figure 1.1 as $C(s) = 1/G(s)$, where $G(s)$ is the plant's transfer function. The design is shown in Figure 1.6.

The total open-loop transfer function from the input is $C(s)G(s) = 1$, which is perfect for tracking the input. What's wrong with that design?

Answer
1. The controller and plant will have different initial conditions; thus, the system will have significant transient response.

2. If the system $G(s)$ is unstable, then due to uncertainty in the system's parameters, the system will still be unstable after adding the controller.

3. Disturbance rejection is very low due to the absence of closed loop.

Conversion from Transfer Function to ODE

Suppose the transfer function is given by $G(s)$ and described by a ratio of two polynomials [numerator $N(s)$ and denominator $D(s)$], that is, $G(s) = \frac{N(s)}{D(s)}$. The transfer function is also defined as a ratio between output $Y(s)$ and input $U(s)$; therefore $G(s) = \frac{N(s)}{D(s)} = \frac{Y(s)}{U(s)}$. We can cross-multiply $N \cdot U$ and $D \cdot Y$. Then, we get the ODE by applying the inverse Laplace transform (Appendix F.10):

$$\mathcal{L}^{-1}\{D(s)Y(s)\} = \mathcal{L}^{-1}\{N(s)U(s)\} \tag{1.11}$$

Example
Given the transfer function $G(s) = \frac{2s+3}{s^2+5s+10}$, the ratio of output to input is defined as $\frac{Y(s)}{U(s)} = \frac{2s+3}{s^2+5s+10}$. We cross-multiply $Y(s)$ with the denominator of the transfer function and $U(s)$ with the numerator of the transfer function and get $(s^2+5s+10)Y(s) =$

$(2s + 3)U(s)$. Now we apply inverse Laplace transform to both sides (by replacing each multiplication by s by a derivative) and get the ODE: $\ddot{y}(t) + 5\dot{y}(t) + 10y(t) = 2\dot{u}(t) + 3u(t)$.

Example
Given the transfer function of DC motor (1.10) and Equation (1.11),

$$\mathcal{L}^{-1}\{(JLs^3 + RJs^2 + K^2s)\Theta(s)\} = \mathcal{L}^{-1}\{KV(s)\}$$

Here s and its powers are used as a derivative operator of the appropriate order working on the appropriate variable; that is, $s^2\Theta(s)$ is converted to $\ddot{\theta}(t)$. Thus,

$$JL\dddot{\theta}(t) + RJ\ddot{\theta}(t) + K^2\dot{\theta}(t) = Kv(t) \tag{1.12}$$

Questions about Equivalence of ODE and Transfer Function Representations
Is Equation (1.12) equivalent to the system of Equations (1.5) and (1.6), $[J\ddot{\theta}(t) = Ki(t)$, and $L\dfrac{di(t)}{dt} + Ri(t) + K\dot{\theta}(t) = v(t)]$, which were the original ODEs describing DC motor?

Can we say that both ODE and transfer function representations are equivalent?

Answer
Yes, Equation (1.12) is equivalent to (1.5) and (1.6). From (1.5) we conclude that $i = \dfrac{J}{K}\ddot{\theta}$ and $\dfrac{di}{dt} = \dfrac{J}{K}\dddot{\theta}$. Now, if we replace i and its derivative in Equation (1.6), we get $\dfrac{LJ}{K}\dddot{\theta} + \dfrac{RJ}{K}\ddot{\theta} + K\dot{\theta} = v$ which is the same ODE as in (1.12).

Despite our result where we were able to get ODE back from the transfer function, an ODE and transfer function representations are not equivalent, because transfer function representation assumes zero initial conditions while ODE representation needs to have initial conditions to be solved.

Stability

There are many different notions of stability. Probably, you have heard about bounded-input bounded-output (BIBO) and asymptotic stability definitions. There are two theorems below which make testing BIBO and asymptotic stability easy for systems represented by a transfer function.

The system is *BIBO stable if and only if* all the system's denominator roots [after the maximal algebraic cancelation (reduction) of numerator and denominator] are "stable." For continuous-time systems, "stable roots" means the roots are in the left semi-plane ($Re\{s\} < 0$). For discrete-time systems, "stable roots" means the roots are inside the unit circle ($|s| < 1$).

The system is *asymptotically stable if and only if* all the system's denominator roots (before the cancelation) are "stable."

NOTE Despite significant similarity between testing criteria for two types of stability, their meaning is very different. BIBO stability means that for any bounded (not diverging) input signal, the output will not diverge. Asymptotic stability means that for input signal 0, starting from any initial conditions, the system will eventually converge to 0.

Example

$$G(s) = \frac{5(s+1)(s-2)}{(s+2)(s+7)(s-2)} = \frac{5(s+1)}{(s+2)(s+7)}$$

The transfer function $G(s)$ is BIBO stable, but not asymptotically stable, because it has denominator roots $-2, -7, 2$ before the cancelation and the unstable root 2 is canceled.

CAUTION! The only difference between asymptotic stability and BIBO stability is in using reduced or not reduced transfer function. Many times, the word "poles" is used for the denominator roots of not reduced transfer function, but this is mathematically incorrect. In the example above, only -2 and -7 are the system's poles.

Example

$$G(s) = \frac{s-3}{s^2-9}$$

A. Check asymptotic and BIBO stability.

The system is not asymptotically stable, because $s^2 - 9 = 0$ has a nonnegative root at 3. It is BIBO stable because after cancelation the transfer function is $G(s) = \frac{1}{s+3}$ with a negative pole at -3.

B. Find the system's ODE representation.

The system is of a second order. We must use the transfer function before cancelation to compute the ODE: $\frac{Y(s)}{U(s)} = \frac{s-3}{s^2-9}$ and $(s^2 - 9)Y(s) = (s-3)U(s)$. By applying inverse Laplace transform to both sides, the obtained ODE is $\ddot{y}(t) - 9y(t) = \dot{u}(t) - 3u(t)$.

C. Compute ZSR for the initial conditions $y(0) = 1$; $\dot{y}(0) = -3$.

For zero input condition $[u(t) = 0]$, the ODE has become homogeneous: $\ddot{y}(t) - 9y(t) = 0$. A general solution of this equation is $y(t) = C_1 e^{-3t} + C_2 e^{3t}$. Now we can substitute initial conditions and solve a system of equations to find the coefficients C_1 and C_2:

$$\begin{cases} y(0) = 1 = C_1 e^0 + C_2 e^0 = C_1 + C_2 \\ \dot{y}(0) = -3 = -3C_1 e^0 + 3C_2 e^0 = -3C_1 + 3C_2 \end{cases}$$

The solution is $C_1 = 1$ and $C_2 = 0$. If we substitute those coefficients back into the expression for $y(t)$, we get $y(t) = C_1 e^{-3t} + C_2 e^{3t} = e^{-3t}$. It is very important to remember that this system is asymptotically unstable, but for those specific initial conditions, the output converges to zero (since $\lim_{t \to \infty} e^{-3t} = 0$). In other words, if the system is converging for some initial conditions, it does not mean that the system is asymptotically stable, because any small disturbance will cause the system to diverge.

Routh-Hurwitz Stability Criterion

Routh (1877) and Hurwitz (1895) independently developed two methods that could be used to identify if some polynomials have roots only in the left semi-plane based on their coefficients. We will not repeat the criterion here, but will only mention that for quadratic and cubic polynomials this criterion is easy to use and remember:

- For a second-degree polynomial to have all stable roots, all its coefficients should be of the same sign (all positive or all negative).
- For a third-degree polynomial $a_3 s^3 + a_2 s^2 + a_1 s + a_0$ to have all stable roots, all its coefficients should be of the same sign and $a_3 a_0 < a_2 a_1$.

What Is So Special about the Left Half-Plane?

If all the system's poles are in the left semi-plane, then the continuous-time system is stable. How does our intuitive notion of stability connect to all these roots and complex semi-planes? What will happen if, for example, all the roots satisfy $Re\{s\} < -2$?

Answer

A general solution of a linear homogeneous ODE $y^{(n)} + a_{n-1} y^{(n-1)} + \cdots + a_1 \dot{y} + a_0 y = 0$ which describes the system (for zero input) has the following form:

$$y(t) = C_1 e^{s_1 t} + C_2 e^{s_2 t} + C_3 e^{s_3 t} + \cdots + C_n e^{s_n t} \tag{1.13}$$

where C_i are constants, and s_i are n complex solutions of $s^n + a_{n-1} s^{n-1} + \cdots + a_1 s + a_0 = 0$.

For real values of s_i, the appropriate exponent in (1.13) is converging to zero with increasing time t only if the power is negative. Thus, s_i must be negative. If the root s_i is complex, that is, $s_i = \alpha + \beta j$, then $e^{s_i t} = e^{(\alpha + \beta j)t} = e^{\alpha t} e^{j \beta t}$. Since $|e^{j\beta t}| = 1$, the exponent will converge only if $\alpha < 0$. Refer to Appendix C for more information. In both cases the convergence condition is $Re\{s_i\} < 0$. For the convergence of $y(t)$, it is required that all exponents converge; thus, all roots should satisfy $Re\{s_i\} < 0$, which is exactly the left semi-plane.

For more negative $Re\{s_i\}$, the exponents $e^{s_i t}$ converge faster to zero, which means that the system becomes faster.

Connection between Performance Specifications in Time and in the Complex Domain

For simple systems, the well-known step response performance specifications (such as overshoot and settling time) could be approximately converted into regions of pole distribution in the complex domain. Using dominant poles approximation, we could assume that the system that has closed-loop poles located inside the trapezoid described in Figure 1.7 is satisfying all the design requirements.

For the second-order system $G(s) = \dfrac{\omega_n^2}{s^2 + 2\zeta \omega_n s + \omega_n^2}$, its settling time and overshoot could be computed by

$$t_{\text{settling}}[2\%] \cong \frac{-\ln(\Delta)}{\zeta \omega_n} \cong \frac{3.9}{\sigma} \tag{1.14}$$

$$O.S.[\%] = 100 e^{-\frac{\zeta \pi}{\sqrt{1-\zeta^2}}} \tag{1.15}$$

10 Chapter One

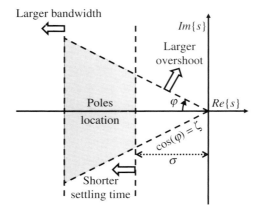

FIGURE 1.7 The connection between performance characteristics of second-order system and location of its poles in the complex domain.

Thus, the larger the distance from the $j\omega$ axis (denoted by $\sigma = \zeta\omega_n$), the shorter the settling time. Also, the larger the angle $\varphi = \arccos(\zeta)$ is, the larger will be the overshoot. It is worth noting that we do not want to put the system's poles too far to the left since this increases the bandwidth of the system and makes it more affected by noises and disturbances.

Question on Controller Design

We want the closed-loop transfer function to be $G_{cl}(s) =$ *some stable rational function*. We know that the closed-loop transfer function with serial controller $C(s)$ equals to $G_{cl}(s) = \dfrac{C(s)G(s)}{1+C(s)G(s)}$, thus by simple algebra [solving for $C(s)$]: $C(s) = \dfrac{G_{cl}(s)}{G(s)(1-G_{cl}(s))}$. What is wrong with that design? Can we design a serial controller $C(s)$ to relocate all the closed-loop poles anywhere we want?

Answer

Sometimes this technique may work well if the closed-loop transfer function has been chosen well. Unfortunately, there are no indicators in this technique for how to choose the "right transfer function." If the choice is bad, we may get in trouble. For example, if the open-loop transfer function is $G(s) = 1/(s-1)$ and the desired closed-loop transfer function is $1/(s+1)$, then the controller will be $C(s) = (s-1)/s$, which means that we have cancelation of the unstable pole. If we have uncertainty (we always have uncertainty in real physical systems) in the location of the plant's pole, say, it is not exactly 1, but close to 1 (such as 1.000001) which causes the transfer function to become $G(s) = 1/(s - 1.000001)$, then the closed-loop system will be $(s - 1)/(s^2 - 0.000001s - 1)$, which is unstable.

State-Space System's Representation

In addition to ODE and transfer function representations, we have another way to describe systems using state-space equations.

State variables are the minimal set of the system's internal signals that can represent, together with the set of inputs, the entire state of the system at any given time.

State space is the space of all possible state variable values.

At first, these definitions might sound too broad and abstract, because they are. Some examples of state variables such as velocity and acceleration, or motor angle, are easy to comprehend, while other examples of variables such as engine air supply divided by a square of its angular velocity make no physical sense. The important thing here is that by using those auxiliary state variables, we can write our original system's ODE as a system of first-order ODEs and this new state-space representation will be completely equivalent to the ODE representation.

We saw in the example on DC motor modeling that the ODE of a given order can be represented by a system of ODEs of lower order. It could be shown that nth-order ODE could be equivalently described by n ODEs of the first order. This is what *state-space representation* is. Instead of the output variable $y(t)$ and its n derivatives in the original ODE, we will have n differential equations of the first order with n state variables that we will denote by $x_1(t), x_2(t), \ldots, x_n(t)$.

NOTE The state variables need not always have a physical meaning, but it is a good idea to choose the state variables from the physical variables (e.g., angle, velocity, current) and their derivatives.

NOTE With the state-space representation, we can't use a serial controller anymore. We need more sophisticated control structures.

NOTE We will frequently skip the time variable (t) to simplify notation.

Why Should We Learn about State Space?

- State-space system representation is equivalent to ODE representation.
- We can relocate the closed-loop poles anywhere we want (given some simple conditions).
- The process of controller design can be easily programmed (it is not a black art anymore).
- Controller design can be simply extended to multiple input multiple output (MIMO) systems.
- Optimal control is based on state-space representation.

Conversion from ODE to State Space

How to choose the state variables is not a trivial question. We will learn some technical ways of conversion later. If the state vector $x = (x_1, x_2, \ldots, x_n)^T$ has been defined in terms of the physical system's variables, we need to compute the first derivative of each x_i as a function of state variables x_j and input u. Then, we compute the output y as a function of the state variables and input. Finally, we present the result in the matrix form: $\dot{x}(t) = Ax(t) + Bu(t); y(t) = Cx(t) + Du(t)$ (continuous-time case), or $x_{n+1} = Ax_n + Bu_n; y_n = Cx_n + Du_n$ (discrete-time case).

Example

A system is represented by the following ODE: $\dddot{y}(t) + 2\ddot{y}(t) + 3y(t) = 4u(t)$. Since the system is of the third order, we need to define three state variables. Presently, it is not clear how to define state variables in terms of y, u, and their derivatives in a way that everything will work out. We will learn about defining state variables later, but now if somebody has defined the variables for us, we should be able to write state-space equations. Let's assume that the state variables were defined by $x_1(t) = y(t); x_2(t) = \dot{y}(t); x_3(t) = \ddot{y}(t)$. Then, we need to compute the first derivatives of all x_i in terms of any elements of a vector x and u. We start with x_1: $\dot{x}_1 = \dot{y}$, thus from the definition of x_2, $\dot{x}_1 = x_2$. Similarly, $\dot{x}_2 = \ddot{y} = x_3$. For the computation of $\dot{x}_3$, we need to use the ODE rearranged in the form $\dddot{y}(t) = -2\ddot{y}(t) - 3y(t) + 4u(t)$. We get $\dot{x}_3 = \dddot{y} = -2\ddot{y} - 3y + 4u = -2x_3 - 3x_1 + 4u$. The system's output is $y = x_1$. Note that we have represented everything in terms of x_i and u. The same equations could be written in matrix form:

$$\begin{pmatrix} \dot{x}_1 \\ \dot{x}_2 \\ \dot{x}_3 \end{pmatrix} = \begin{pmatrix} 0 & 1 & 0 \\ 0 & 0 & 1 \\ -3 & 0 & -2 \end{pmatrix} \begin{pmatrix} x_1 \\ x_2 \\ x_3 \end{pmatrix} + \begin{pmatrix} 0 \\ 0 \\ 4 \end{pmatrix} u$$

$$y = \begin{pmatrix} 1 & 0 & 0 \end{pmatrix} \begin{pmatrix} x_1 \\ x_2 \\ x_3 \end{pmatrix} + 0 \cdot u$$

Example

We return to Equations (1.5) and (1.6):

$$J\ddot{\theta}(t) = Ki(t) \qquad (1.5)$$

$$L\frac{di(t)}{dt} + Ri(t) + K\dot{\theta}(t) = v(t) \qquad (1.6)$$

Let's define the state as follows: $x_1 = i; x_2 = \theta; x_3 = \dot{\theta}$. Then, by definition $\dot{x}_1 = \frac{di}{dt}$ and from (1.6):

$$\dot{x}_1 = \frac{di}{dt} = \frac{1}{L}(v - Ri - K\dot{\theta}) = \frac{1}{L}u - \frac{R}{L}x_1 - \frac{K}{L}x_3$$

$$\dot{x}_2 = \dot{\theta} = x_3$$

By definition, $\dot{x}_3 = \ddot{\theta}$ and from (1.5):

$$\dot{x}_3 = \ddot{\theta} = \frac{Ki}{J} = \frac{Kx_1}{J}$$

$$y = \theta = x_2$$

or, in short,

$$\dot{x} = \begin{pmatrix} -\frac{R}{L} & 0 & -\frac{K}{L} \\ 0 & 0 & 1 \\ \frac{K}{J} & 0 & 0 \end{pmatrix} x + \begin{pmatrix} \frac{1}{L} \\ 0 \\ 0 \end{pmatrix} u$$

$$y = (0 \quad 1 \quad 0)x + 0u$$

CAUTION! It is advisable at this point to review the mathematical tools provided in Appendices F and G to make sure that you can find those tools later when you need them.

Solved Problems

Problem 1.1
The state-space system is given by

$$\begin{cases} \begin{pmatrix} \dot{x}_1 \\ \dot{x}_2 \end{pmatrix} = \begin{pmatrix} -1 & -1 \\ 0 & -1 \end{pmatrix} \begin{pmatrix} x_1 \\ x_2 \end{pmatrix} + \begin{pmatrix} 5 \\ 10 \end{pmatrix} u \\ y = (-1 \quad 1) \begin{pmatrix} x_1 \\ x_2 \end{pmatrix} \end{cases}$$

Compute $x_1(t)$, $x_2(t)$, and $y(t)$ for $t > 0$, if the initial conditions are $x_1(0) = 0$ and $x_2(0) = 1$ and the input is $u(t) = 0$.

Solution
We can write the equations explicitly by multiplying matrices and vectors:

$$\begin{cases} \dot{x}_1 = -x_1 - x_2 + 5u \\ \dot{x}_2 = -x_2 + 10u \\ y = -x_1 + x_2 \end{cases}$$

When $u = 0$, the second equation becomes $\dot{x}_2 = -x_2$ with obvious solution $x_2(t) = C_2 e^{-t}$ and some constant C_2. Since $x_2(0) = 1$, the constant must be $C_2 = 1$, thus $x_2(t) = e^{-t}$. The first differential equation becomes $\dot{x}_1 = -x_1 - x_2 = -x_1 - e^{-t}$. We use the formula from Appendix F.11 to solve this ODE [$p(t) = 1$ and $q(t) = -e^{-t}$]:

$$x_1(t) = \exp\left(-\int_0^t p(\tau)d\tau\right)\left[x_1(0) + \int_0^t q(\tau)\exp\left(\int_0^\tau p(\sigma)d\sigma\right)d\tau\right] = e^{-t}\left(0 + \int_0^t -e^{-\tau}e^{\tau}d\tau\right) = -te^{-t}$$

Finally, $y(t) = -x_1 + x_2 = (t+1)e^{-t}$.

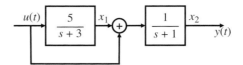

FIGURE 1.8 Open-loop system block diagram.

Problem 1.2
The system block diagram and the definition of state variables are given in Figure 1.8. Write state-space equations of this system.

Solution
From the left transfer function:

$$\frac{X_1(s)}{U(s)} = \frac{5}{s+3} \Rightarrow \dot{x}_1 + 3x_1 = 5u \Rightarrow \dot{x}_1 = -3x_1 + 5u$$

From the right transfer function (with the input $x_1 + u$):

$$\frac{X_2(s)}{X_1(s)+U(s)} = \frac{1}{s+1} \Rightarrow \dot{x}_2 + x_2 = x_1 + u \Rightarrow \dot{x}_2 = x_1 - x_2 + u$$

and $y = x_2$.

Now, the same equations could be written in matrix form:

$$\dot{x} = \begin{pmatrix} -3 & 0 \\ 1 & -1 \end{pmatrix} x + \begin{pmatrix} 5 \\ 1 \end{pmatrix} u$$

$$y = (0 \quad 1)x$$

where $x = (x_1 \quad x_2)^T$.

Problem 1.3
The system's block diagram is given in Figure 1.9.
Write the state-space equations $\begin{cases} \dot{x} = Ax + Bu \\ y = Cx \end{cases}$ for this system when the state variables are defined as follows:

$$\begin{cases} x_1 = w \\ x_2 = \dot{y} - e \\ x_3 = y \end{cases} \quad (1.16)$$

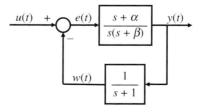

FIGURE 1.9 Closed-loop system block diagram.

Solution
First, we will write everything we know on the system from simple observations. Remember that the goal is to have all derivatives on the left side and state variables and inputs on the right side of equations. Based on state variable definitions,

$$\begin{cases} \dot{x}_1 = \dot{w} \\ \dot{x}_2 = \ddot{y} - \dot{e} \\ \dot{x}_3 = \dot{y} \end{cases} \quad (1.17)$$

We also can see that

$$\frac{Y(s)}{E(s)} = \frac{s+\alpha}{s(s+\beta)} \Rightarrow \ddot{y} + \beta\dot{y} = \dot{e} + \alpha e \quad (1.18)$$

$$\frac{W(s)}{Y(s)} = \frac{1}{s+1} \Rightarrow \dot{w} + w = y \quad (1.19)$$

$$e = u - w \quad (1.20)$$

Now, we are ready to write state-space equations. From (1.16), (1.17), and (1.19)

$$\dot{x}_1 = \dot{w} = y - w = x_3 - x_1$$

From (1.17), (1.18), and (1.20)

$$\dot{x}_2 = \ddot{y} - \dot{e} = -\beta\dot{y} + \alpha e = -\beta(\dot{y} - e) - \beta e + \alpha e = -\beta x_2 + e(\alpha - \beta)$$
$$= -\beta x_2 + (u - w)(\alpha - \beta) = -\beta x_2 + (u - x_1)(\alpha - \beta)$$

From (1.16), (1.17), and (1.20)

$$\dot{x}_3 = \dot{y} = x_2 + e = x_2 + u - w = x_2 - x_1 + u$$

Alternatively, we could use the matrix form

$$\dot{x} = \begin{pmatrix} -1 & 0 & 1 \\ \beta - \alpha & -\beta & 0 \\ -1 & 1 & 0 \end{pmatrix} x + \begin{pmatrix} 0 \\ \alpha - \beta \\ 1 \end{pmatrix} u$$

The equation for the output y is already given in (1.16), thus $y = x_3 = (0 \quad 0 \quad 1)x$.

Problem 1.4
For the system given in Figure 1.9, what is the condition on α and β that the closed-loop system will be stable (using Routh-Hurwitz criterion)? Draw approximately the region of stability in (α, β) system of coordinates.

Solution
To be stable, the denominator of the closed-loop transfer function should have all roots in the left semi-plane. The closed-loop transfer function is

$$G_{cl}(s) = \frac{\frac{s+\alpha}{s(s+\beta)}}{1 + \frac{s+\alpha}{(s(s+\beta)(s+1))}} = \frac{(s+1)(s+\alpha)}{s^3 + (\beta+1)s^2 + (\beta+1)s + \alpha}$$

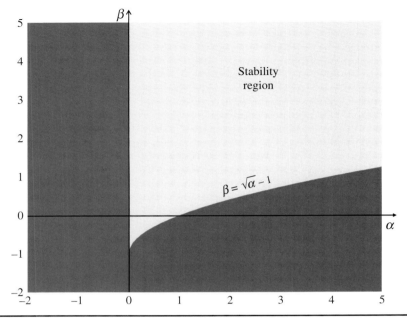

Figure 1.10 Stability region for a system described in Figure 1.9.

So, using Routh-Hurwitz criterion we need to test the polynomial $s^3 + (\beta+1)s^2 + (\beta+1)s + \alpha$. The cubic polynomial is stable if all its coefficients are positive and the multiplication of two middle coefficients is greater than the multiplication of two other coefficients. In other words, the stability will be achieved for $\alpha > 0$, $\beta + 1 > 0$, and $(\beta+1)^2 > \alpha$. Since $\alpha > 0$, the last inequality could be simplified to $\beta > \sqrt{\alpha} - 1$. The intersection of all these conditions is given in Figure 1.10.

CHAPTER 2
State-Space Representations

In this chapter, we will learn about the linear systems theory with a general state-space representation, and transformations from transfer functions to state space and back. We will also learn the stability criteria for continuous- and discrete-time systems represented by state-space equations. The terms controllability, observability, and minimality will also be introduced. It is important to get used to the notation introduced in this chapter since it will be used in all subsequent chapters. It will be quite useful to brush up your knowledge of the Laplace transform and linear algebra using Appendices F and G before you start reading this chapter.

Continuous-Time Single-Input Single-Output (SISO) State-Space Systems

A general *linear time-invariant* (LTI) state-space system (see Appendix F) is given by

$$\begin{cases} \dot{x}_{n\times 1}(t) = A_{n\times n} x_{n\times 1}(t) + B_{n\times 1} u_{1\times 1}(t) \\ y_{1\times 1}(t) = C_{1\times n} x_{n\times 1}(t) + D_{1\times 1} u_{1\times 1}(t) \\ x(0) = x_0 \end{cases} \qquad (2.1)$$

where $x(t) \in \mathbb{R}^n$ is the real n-dimensional state vector $x(t) = (x_1(t), x_2(t), \ldots, x_n(t))^T$; and $u(t)$ and $y(t)$ are the scalar input and output respectively. When you start learning about state-space equations, it is critical that you remember the dimensions of matrices and vectors involved in (2.1). Note that x and B are column vectors, A is a square matrix, C is a row vector, and D is a scalar. So, more explicitly, the system equations could be written as (while we deliberately skip adding time t to simplify notation):

$$\begin{cases} \begin{pmatrix} \dot{x}_1 \\ \dot{x}_2 \\ \vdots \\ \dot{x}_n \end{pmatrix} = \begin{pmatrix} a_{11} & a_{12} & \cdots & a_{1n} \\ a_{21} & a_{22} & \cdots & a_{2n} \\ \vdots & \vdots & \ddots & \vdots \\ a_{n1} & a_{n2} & \cdots & a_{nn} \end{pmatrix} \begin{pmatrix} x_1 \\ x_2 \\ \vdots \\ x_n \end{pmatrix} + \begin{pmatrix} b_1 \\ b_2 \\ \vdots \\ b_n \end{pmatrix} u \\ y = (c_1 \quad c_2 \quad \cdots \quad c_n) \begin{pmatrix} x_1 \\ x_2 \\ \vdots \\ x_n \end{pmatrix} + du \end{cases} \quad (2.2)$$

where $a_{ij}, b_i, c_i,$ and d are elements of matrices A, B, C, D.

NOTE Since the structure of the system's representation is constant and completely defined by the matrices and vectors A, B, C, D, sometimes we will define a system by just giving its matrices $\{A, B, C, D\}$ or even $\{A, B, C\}$ if $D = 0$, instead of writing (2.2) explicitly.

Transfer Function of Continuous-Time SISO State-Space System

To find the transfer function of a state-space system, we can apply the Laplace transform to both sides of state Equations (2.1) and (2.2). The goal is to find the ratio $Y(s)/U(s)$, where $Y(s)$ is the Laplace transform of output $y(t)$ and $U(s)$ is the Laplace transform of input $u(t)$. Note that the Laplace transform of a derivative of $f(t)$ is $sF(s) - f(0)$, and applying the Laplace transform to vectors is just applying it to each element of the vector separately. Thus,

$$\begin{cases} sX(s) - x_0 = AX(s) + BU(s) \\ Y(s) = CX(s) + DU(s) \end{cases} \quad (2.3)$$

where $x_0 = x(0)$ is the vector of state variable initial conditions.

When moving $AX(s)$ to the left side and x_0 to the right side in the first equation of (2.3), we get $(sI - A)X(s) = x_0 + BU(s)$. To get rid of $(sI - A)$, we multiply both sides from the left by $(sI - A)^{-1}$ and get

$$X(s) = (sI - A)^{-1} x_0 + (sI - A)^{-1} BU(s) \quad (2.4)$$

NOTE If we apply the inverse Laplace transform to both sides of (2.4), we could get an explicit solution for $x(t) = e^{At} x_0 + e^{At} * Bu(t)$, where $*$ denotes convolution (see Appendix F.4).

Now, we can substitute $X(s)$ from (2.4) into the second equation of (2.3) and get

$$Y(s) = C(sI - A)^{-1} x_0 + C(sI - A)^{-1} BU(s) + DU(s) \quad (2.5)$$

The $C(sI - A)^{-1} x_0$ is *zero input response* (ZIR), and $C(sI - A)^{-1} BU(s) + DU(s)$ is *zero state response* (ZSR).

State-Space Representations

Thus, by dividing both sides of (2.5) by $U(s)$ and assuming $x_0 = 0$, the transfer function which is defined by the output to input ratio (for zero initial conditions) is

$$G(s) = \left.\frac{Y(s)}{U(s)}\right|_{x_0=0} = C(sI - A)^{-1}B + D \tag{2.6}$$

Alternatively, it is sometimes more convenient to use (2.7) or (2.8) to compute transfer functions.

$$G(s) = \frac{\det\begin{pmatrix} sI - A & -B \\ C & D \end{pmatrix}}{\det(sI - A)} \tag{2.7}$$

$$G(s) = \frac{\det(sI - A + BC)}{\det(sI - A)} + D - 1 \tag{2.8}$$

All three formulas are equivalent. For example, to derive (2.6) from (2.7), we could use the determinant of the block matrices formula (see Appendix G.21):

$$\frac{\det\begin{pmatrix} sI - A & -B \\ C & D \end{pmatrix}}{\det(sI - A)} = \frac{\det(sI - A)\det(D - C(sI - A)^{-1}(-B))}{\det(sI - A)} = \det(C(sI - A)^{-1}B + D)$$

So, we have got something very similar to (2.6), but with a determinant. Why are both formulas the same then? Note that the expression inside the determinant is a scalar polynomial; thus, its determinant is the polynomial itself, which shows that (2.6) and (2.7) are equivalent.

NOTE The transfer function is always of the order n (or less if there are cancelations in the numerator and denominator polynomials) if matrix A has dimensions $n \times n$.

NOTE Formula (2.7) has a very interesting form. Both determinants are producing polynomials and the ratio of those polynomials is the transfer function, which implies that the *characteristic polynomial* (polynomial of poles) is $\det(sI - A)$ and the polynomial of the system's zeros is $\det\begin{pmatrix} sI - A & -B \\ C & D \end{pmatrix}$. To find poles and zeros of the system, we need to equate those corresponding determinants to 0 and solve the polynomial equations.

Question
When is it more convenient to use different formulas for computing transfer functions, that is, compare (2.6) versus (2.7) versus (2.8)?

Answer
Equation (2.8) is obviously more convenient if $D = 1$. For low-order systems, especially when using hand computations, Formula (2.7) is more convenient because using inverting matrix is a cumbersome process. Formula (2.6) is used in other cases.

It is possible to compute the explicit solution of matrix differential Equation (2.2) by using the inverse Laplace transform of (2.4) and (2.5):

$$x(t) = e^{At}x_0 + \int_0^t e^{A(t-\tau)}Bu(\tau)d\tau \qquad (2.9)$$

$$y(t) = Ce^{At}x_0 + \int_0^t Ce^{A(t-\tau)}Bu(\tau)d\tau + Du(t) \qquad (2.10)$$

The matrix e^{At} is very important in control theory, and is called *transition matrix*. To understand what this matrix is, how it is computed, and how we get to (2.9) and (2.10), refer to Appendix G.12. Though (2.9) and (2.10) are rarely used in computations or for solving problems, they are important from the theoretical standpoint because some of the developments that we will be discussing are explained by those formulas.

Question
How is it possible (if at all) to realize a transfer function $H(s) = \alpha s + \beta$, where α and β are constants (ideal PD [proportional derivative] controllers) in state space?

Answer
The function includes pure derivative; thus, it is noncausal. Theoretically, it is not possible to implement noncausal state-space system, though practical ways exist to overcome this limitation.

Discrete-Time SISO State-Space Systems

A general LTI state-space system in discrete time is given by

$$\begin{cases} x_{n\times 1}[k+1] = A_{n\times n}x_{n\times 1}[k] + B_{n\times 1}u_{1\times 1}[k] \\ y_{1\times 1}[k] = C_{1\times n}x_{n\times 1}[k] + D_{1\times 1}u_{1\times 1}[k] \\ x[0] = x_0 \end{cases} \qquad (2.11)$$

where $k = 0, 1, 2, \ldots$ is sample number.

An explicit solution of system (2.11) is given by

$$x[k] = A^k x_0 + \sum_{m=0}^{k-1} A^{k-1-m}Bu[m] = A^k x_0 + \sum_{m=0}^{k-1} A^m Bu[k-m-1] \qquad (2.12)$$

$$y[k] = CA^k x_0 + \left(\sum_{m=0}^{k-1} CA^m Bu[k-m-1]\right) + Du[k] \qquad (2.13)$$

Transfer Function of Discrete-Time SISO State-Space System

Similar to (2.6), by applying Z-transform (see Appendix F.12) we obtain

$$G(z) = \frac{Y(z)}{U(z)} = C(zI-A)^{-1}B + D = \frac{\det\begin{pmatrix} zI-A & -B \\ C & D \end{pmatrix}}{\det(zI-A)} = \frac{\det(zI-A+BC)}{\det(zI-A)} + D - 1 \qquad (2.14)$$

Multiple-Input Multiple-Output (MIMO) State-Space Systems

The LTI MIMO state-space system with m inputs and l outputs is given by

$$\begin{cases} \dot{x}_{n\times 1}(t) = A_{n\times n} x_{n\times 1}(t) + B_{n\times m} u_{m\times 1}(t) \\ y_{l\times 1}(t) = C_{l\times n} x_{n\times 1}(t) + D_{l\times m} u_{m\times 1}(t) \\ x(0) = x_0 \end{cases} \quad (2.15)$$

where $u(t) = (u_1(t), \ldots, u_m(t))^T$ are inputs and $y(t) = (y_1(t), \ldots, y_l(t))^T$ are outputs.

NOTE Most of the formulas we develop in this course are appropriate for MIMO systems without any significant change (or without any change).

NOTE The state equations can be written more explicitly

$$\begin{cases} \dot{x} = Ax + B_1 u_1 + \cdots + B_m u_m \\ y_j = C_j x + D_j u; \quad j = 1, \ldots, l \\ x(0) = x_0 \end{cases} \quad (2.16)$$

where B_i are the columns of B, and C_j and D_j are the rows of C and D matrices, respectively.

Now, the transfer function $G(s) = C(sI - A)^{-1}B + D$ is an $l \times m$ matrix, where

$$[G(s)]_{ij} = \frac{Y_i(s)}{U_j(s)} \quad (2.17)$$

CAUTION! Even if all $[G(s)]_{ij}$ are stable, the MIMO system is not necessarily stable.

Stability of Continuous-Time Systems

If we return to Formula (2.7), we may see that the denominator of the transfer function is the *characteristic polynomial* of the matrix A (roots of a characteristic polynomial are the eigenvalues of A). Therefore, the asymptotic stability of the system can be determined using the eigenvalues of A (the system's poles are always a subset of A's eigenvalues).

Theorem 2.1
The system is asymptotically stable if and only if all eigenvalues of A have negative real part ($\forall s_i, i = 1, \ldots, n$: $Re\{s_i\} < 0$) where $\forall$ symbol denotes "for all… ."

Stability of Discrete-Time Systems

For discrete-time system, the asymptotic stability is verified using the following theorem.

Theorem 2.2
The system is asymptotically stable if and only if all eigenvalues of A are inside the unit circle ($\forall s_i, i = 1, \ldots, n$: $|s_i| < 1$).

Example

The system is given by

$$\begin{cases} \dot{x}(t) = \begin{pmatrix} 0 & 9 \\ 1 & 0 \end{pmatrix} x(t) + \begin{pmatrix} 1 \\ 0 \end{pmatrix} u(t) \\ y(t) = (1 \quad -3) x(t) \end{cases} \quad (2.18)$$

A. Check the asymptotic stability.

To check the asymptotic stability, we need to find eigenvalues of the matrix $\begin{pmatrix} 0 & 9 \\ 1 & 0 \end{pmatrix}$. Those eigenvalues are the solutions of the following polynomial equation: $\det(sI - A) = \det\left(s\begin{pmatrix} 1 & 0 \\ 0 & 1 \end{pmatrix} - \begin{pmatrix} 0 & 9 \\ 1 & 0 \end{pmatrix}\right) = 0$. Thus, $\begin{vmatrix} s & -9 \\ -1 & s \end{vmatrix} = s^2 - 9 = 0$.

The eigenvalues are 3 and -3. The first eigenvalue is positive; therefore, the system is not asymptotically stable.

B. Compute the transfer function and check the BIBO stability.

To compute the transfer function, we will use (2.7): $G(s) = \dfrac{\det\begin{pmatrix} sI - A & -B \\ C & D \end{pmatrix}}{\det(sI - A)} =$

$\dfrac{\det\begin{pmatrix} s & -9 & -1 \\ -1 & s & 0 \\ 1 & -3 & 0 \end{pmatrix}}{s^2 - 9} = \dfrac{-1\begin{vmatrix} -1 & s \\ 1 & -3 \end{vmatrix}}{s^2 - 9} = \dfrac{s-3}{s^2 - 9} = \dfrac{1}{s+3}$. After cancelation, the transfer function has a pole at -3; thus, the system is BIBO stable.

Block Diagrams of State-Space Systems

It is easy to draw a block diagram of the state-space system when you know the process. This block diagram generally includes only three basic elements: integrator, gain, and adder (or subtractor).

NOTE In the Laplace domain, integration is produced by division by s, or multiplying by $\dfrac{1}{s}$.

Any LTI state-space system could be built from these basic elements. The built model could be used in simulations, or for a better understanding of the system's interconnections.

To draw a block diagram, the following process should be followed:

1. Identify the system's order n and draw n integrator boxes (with input and output arrows) with enough space between the blocks.

2. For integrator i ($i = 1, \ldots, n$), the input of integrator is denoted by $\dot{x}_i$ and the output by x_i.

3. Draw an input arrow $u(t)$ from the left side of a page and an output arrow $y(t)$ on the right side.

4. Write all state differential equations explicitly as a system of n equations of the kind $\dot{x}_i = \alpha x_1 + \beta x_2 + \cdots + \gamma x_n + \zeta u$.

5. Implement each equation $i (i = 1, \ldots, n)$ by using gains and adders, starting by taking all the x_j and u that appear in the equation and then multiplying them by the appropriate gain (their coefficient in the equation). You could use negative gains or subtractors if needed to add them all together and return to the input of the integrator i. For example, to implement the equation $\dot{x}_2 = 2x_2 - 3x_3 + 5u$, you need to multiply the input by 5, the output of the second integrator by 2 and the output of the third integrator by -3 and then add them all. The sum arrow will be connected to the input of the second integrator.

6. Finally, you need to write the equation for $y(t)$ explicitly and implement it as before (just with the sum arrow connected to the $y(t)$ output arrow).

Example

Let's return to a previous example and draw a block diagram of (2.18) to clarify the process. The system is of the second order.

Figure 2.1 demonstrates the process of implementing these equations step by step.

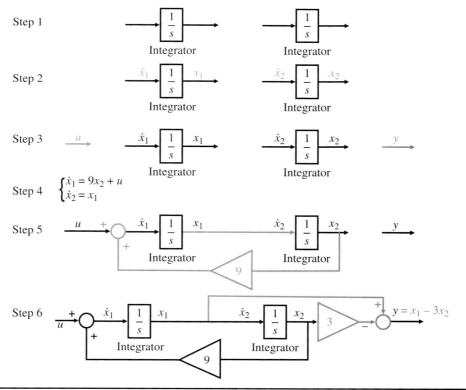

FIGURE 2.1 Block diagram of the state-space system drawing process.

Controllability

The state-space representation of an LTI system is *controllable* if the state vector x can be brought from any initial state x_0 to any final state x_f in finite time, using the appropriate inputs u.

Theorem 2.3
1. The system of order n is controllable if and only if rank($\mathcal{C}$) = n, where $\mathcal{C}$ = $(B, AB, A^2B, \ldots, A^{n-1}B)$ is a *controllability matrix*.

2. The system's state may move only in the directions spanned by the columns of $\mathcal{C}$ (controllable subspace).

NOTE For SISO systems, the matrix $\mathcal{C}$ has the dimensions $n \times n$; for MIMO systems it has the dimensions $n \times mn$, where m is the number of inputs.

NOTE It is easier to understand where the matrix $\mathcal{C}$ comes from in discrete-time case when examining Equation (2.12). For zero initial conditions, the only states the system could get are defined by a linear combination of $\mathcal{C}$ columns. The expression $A^m Bu[k-m-1]$ inside the sum is essentially columns of the matrix $\mathcal{C}$ multiplied by various scalar inputs u. Since we choose the inputs, we could pick any numbers for u that will serve as coefficients of $\mathcal{C}$ columns. Thus, any obtainable state x is the initial vector x_0 multiplied by n's power of A plus a linear combination of $\mathcal{C}$ columns. For example, if all columns are proportional, then the state of the system will be $x[n] = A^n x_0 + \text{const} \cdot B$ since all combinations of $\mathcal{C}$ columns will be proportional to B.

NOTE The input u such that the system will move from state x_0 to state x_f immediately for a continuous-time system or in n steps for a discrete-time controllable system is

$$u(t) = \sum_{m=0}^{n-1} \mathcal{C}^{-1}(e^{-At_f} x_f - x_0) \delta^{(m)}(t) \quad \text{(continuous-time)} \quad (2.19)$$

$$u[k] = \begin{pmatrix} u[n-1] \\ \vdots \\ u[0] \end{pmatrix} = \mathcal{C}^{-1}(x_f - A^n x_0) \quad \text{(discrete-time)} \quad (2.20)$$

In discrete-time case, at the most n steps are necessary to get from any state to any other state. The vector of inputs $u[k]$ in Formula (2.20) is sorted bottom up starting with $u[0]$ at the bottom.

Question
Will the same controllability property work if we change the definition of $\mathcal{C}$ to $\mathcal{C}^T$? Or to backward order of blocks in $\mathcal{C}$: $(A^{n-1}B, \ldots, B)$?

Answer
Since row and column spaces have the same rank, the transposed matrix has the same rank and the controllability property is not changed. Similarly, the order of columns does not matter for rank computations; thus, the controllability is not changed.

Question
How can we check controllability of a MIMO system ($\mathcal{C}$ is not square)?

Answer
The procedure is the same: We need to check if the rank of $\mathcal{C}$ equals to the system's order. In the SISO case, we might use invertibility properties of $\mathcal{C}$ matrix such as $\det(\mathcal{C}) \neq 0$ for full rank. Unfortunately, for MIMO systems, we do not have such shortcuts.

Observability

The state-space representation of the LTI system is *observable* if it is possible to compute the initial state x_0 from the measurement of inputs u and outputs y in a finite time interval (given the matrices A, B, C, D).

NOTE If we can compute the initial state, we can compute the state at any given point of time.

Theorem 2.4

1. The system of order n is observable if and only if rank($\mathcal{O}$) = n, where $\mathcal{O} = \begin{pmatrix} C \\ CA \\ \vdots \\ CA^{n-1} \end{pmatrix}$ is an *observability matrix*.
2. The subspace of states x_0 that can be reconstructed from the inputs and outputs is called *observability subspace*, and it is a subspace of $\mathcal{O}$ rows.

NOTE To compute the controllability or observability matrix, there is no need to compute all the powers of A matrix up to $n-1$. All you need for controllability is to compute AB, then A multiplied by the vector (AB), which is A^2B; then again A multiplied by the vector A^2B, etc. Similarly, for observability, multiply C, CA, etc., each time by A (from the right side).

Minimal Systems

The SISO system that cannot be implemented with a smaller-order state-space realization is called minimal. Such systems have no cancelation in the numerator and denominator of the transfer function $C(sI - A)^{-1}B + D$.

Theorem 2.5
The system is minimal if and only if the system is controllable and observable.

NOTE If there is a cancelation in the transfer function, then the state-space realization is not controllable, or not observable, or both.

State Similarity Transforms

Any given state-space realization can be converted into another state-space realization of the same system by applying a *similarity transform* (two realizations, the original and the transformed are called *similar*, see Appendix G.19; both have the same transfer function).

Let's say the system is defined by state matrices A, B, C, D. We can define new state variables $\tilde{x}$ in terms of the original state variables x, using an arbitrary regular (invertible) *similarity matrix T*:

$$\tilde{x} = T^{-1}x \quad \text{(equivalently } x = T\tilde{x}) \tag{2.21}$$

Now, given the state-space system $\begin{cases} \dot{x} = Ax + Bu \\ y = Cx + Du \end{cases}$, if we replace x with $T\tilde{x}$ everywhere, we get $\begin{cases} T\dot{\tilde{x}} = AT\tilde{x} + Bu \\ y = CT\tilde{x} + Du \end{cases}$. By multiplying the left and right sides of the top equation by T^{-1},

we get $\begin{cases} \dot{\tilde{x}} = \underbrace{T^{-1}AT}_{\tilde{A}}\tilde{x} + \underbrace{T^{-1}B}_{\tilde{B}}u \\ y = \underbrace{CT}_{\tilde{C}}\tilde{x} + \underbrace{D}_{\tilde{D}}u \end{cases}$. Note that the form in which the new system is given is still the same standard state-space form, but in terms of $\tilde{x}$ and the same input and output. Multiplication by T or T^{-1} is not changing matrix and vector dimensions.

It follows that the similar system realization is

$$\tilde{A} = T^{-1}AT; \quad \tilde{B} = T^{-1}B; \quad \tilde{C} = CT; \quad \tilde{D} = D \tag{2.22}$$

$$\tilde{\mathcal{C}} = T^{-1}\mathcal{C}; \quad \tilde{\mathcal{O}} = \mathcal{O}T \tag{2.23}$$

The formulas in (2.23) are easily derived by noticing that $\tilde{A}^k = \underbrace{(T^{-1}AT)(T^{-1}AT)(T^{-1}AT)\cdots(T^{-1}AT)}_{k \text{ times}} = T^{-1}A^kT$, thus $\tilde{\mathcal{C}} = (\tilde{B} \quad \tilde{A}\tilde{B} \quad \tilde{A}^2\tilde{B} \quad \cdots) = (T^{-1}B$

$T^{-1}ATT^{-1}B \quad T^{-1}A^2TT^{-1}B \quad \cdots) = T^{-1}(B \quad AB \quad A^2B \quad \cdots) = T^{-1}\mathcal{C}$. Similarly, $\tilde{\mathcal{O}} = \begin{pmatrix} \tilde{C} \\ \tilde{C}\tilde{A} \\ \tilde{C}\tilde{A}^2 \\ \vdots \end{pmatrix} =$

$\begin{pmatrix} CT \\ CTT^{-1}AT \\ CTT^{-1}A^2T \\ \vdots \end{pmatrix} = \begin{pmatrix} C \\ CA \\ CA^2 \\ \vdots \end{pmatrix}T = \mathcal{O}T$.

Similar realizations have

1. The same eigenvalues
2. The same transfer function
3. The same controllability and observability properties (rank($\mathcal{C}$) = rank($\tilde{\mathcal{C}}$) and rank($\mathcal{O}$) = rank($\tilde{\mathcal{O}}$)).

These properties are easily proven using linear algebra:

1. We need to prove that A and $T^{-1}AT$ have the same characteristic polynomial: $\det(sI - A) = \det(sI - T^{-1}AT)$. Let's start with the right part: $\det(sI - T^{-1}AT) = \det(T^{-1}(TsIT^{-1} - TT^{-1}ATT^{-1})T) = \det(T^{-1}(sI - A)T) = \cancel{\det(T^{-1})}\det(sI - A)\cancel{\det(T)} = \det(sI - A)$ as expected.

2. The transfer function for the transformed system is $\tilde{G}(s) = \tilde{C}(sI - \tilde{A})^{-1}\tilde{B} + \tilde{D} = CT(sI - T^{-1}AT)^{-1}T^{-1}B + D = C(T(sI - T^{-1}AT)T^{-1})^{-1}B + D = C(sI_A)^{-1}B + D = G(s)$. Note that when entering T and T^{-1} into the parentheses, the order of matrices is changing.

3. As shown above, the controllability matrix of the transformed system is $T^{-1}\mathcal{C}$. Thus, the rank of the transformed controllability matrix is the same (not changed by the multiplication by invertible matrix).

Question
How many state-space realizations exist for a given system?

Answer
Since any invertible matrix T generates a new realization, the number of possible similar realizations is infinite.

Question
If the system's realization is not controllable, does it mean that all other realizations of this system are not controllable?

Answer
No, a system that is not controllable might have controllable realization, but this realization will not be observable. Also, you will not be able to find a similarity transform between two realizations.

Canonical Forms

Now that we are ready to convert transfer functions into a state-space form the easy way, we will demonstrate three different realizations that could be obtained automatically from the transfer function.

NOTE A *strictly proper system* has more poles than zeros, and a *proper system* could have an equal number of poles and zeros (or less zeros). The implementation is available only for proper systems (or strictly proper systems). This makes sense since such systems are causal, and we don't need future measurements to compute present time values of a proper system.

Suppose the transfer function is given by

$$G(s) = \frac{b_1 s^{n-1} + \cdots + b_n}{s^n + a_1 s^{n-1} + \cdots + a_n} + d$$

The *controller (controllable) canonical form* is given by

$$\dot{x} = \underbrace{\begin{pmatrix} -a_1 & -a_2 & \cdots & -a_n \\ 1 & 0 & \cdots & 0 \\ & \ddots & & \vdots \\ 0 & & 1 & 0 \end{pmatrix}}_{A_c} x + \underbrace{\begin{pmatrix} 1 \\ 0 \\ \vdots \\ 0 \end{pmatrix}}_{B_c} u; \quad y = \underbrace{(b_1 \ b_2 \ \cdots \ b_n)}_{C_c} x; \quad D_c = d \quad (2.24)$$

In this form, the top row of A_c consists of negative coefficients of the denominator polynomial (excluding the coefficient of s^n), the diagonal below the main diagonal consists of 1s, and everything else is 0. The vector B_c has 1 at the top row and zeros elsewhere, and the vector C_c has numerator coefficients.

Example

The transfer function is given by $G(s) = \dfrac{3}{s^2+2} = \dfrac{b_1 s^{n-1}+\cdots+b_n}{s^n + a_1 s^{n-1}+\cdots+a_n} + d$. Obviously, the system is of the order $n = 2$ (the highest power of s in the transfer function). By comparing coefficients, we get $b_1 = 0$; $b_2 = 3$; $a_1 = 0$; $a_2 = 2$; $d = 0$. So, the matrices will be

$$A_c = \begin{pmatrix} 0 & -2 \\ 1 & 0 \end{pmatrix}; \quad B_c = \begin{pmatrix} 1 \\ 0 \end{pmatrix}; \quad C_c = (0 \;\; 3); \quad D_c = 0$$

NOTE The matrix A_c is given in a *companion form*.

Theorem 2.6

The system in a controller canonical form is always controllable, and its controllability matrix satisfies

$$\mathcal{C}_c^{-1} = \begin{pmatrix} 1 & a_1 & \cdots & a_n \\ 0 & \ddots & \ddots & \vdots \\ \vdots & \ddots & \ddots & a_1 \\ 0 & \cdots & 0 & 1 \end{pmatrix} \quad (2.25)$$

The system realization is controllable if and only if the realization can be transformed to a controller canonical form. The appropriate transformation is

$$T = \mathcal{C}\mathcal{C}_c^{-1} \quad (2.26)$$

NOTE If the system is proper, but not strictly proper, then the D value will not be 0. The rational transfer function should be brought to a form which includes strictly proper transfer function plus a constant. One way to do that is by long division of numerator and denominator, but this is not the easiest way. There is one simple trick that could help as in the following example.

Example

We have to realize the transfer function $G(s) = \dfrac{2s^2+3s+5}{s^2+s+1}$. This would not fit into the transfer function shape used in the canonical form. Instead of dividing $2s^2+3s+5$ by s^2+s+1, we can get rid of s^2 in the numerator by bringing the numerator to a more convenient form by completing it to twice the denominator $2(s^2+s+1)$: $G(s) = \dfrac{2(s^2+s+1)+s+3}{s^2+s+1}$. Now we can cancel and get $G(s) = 2 + \dfrac{s+3}{s^2+s+1}$, where $d = 2$ and the rest is written in the appropriate form.

The *observer (observable) canonical form* for the same transfer function (2.24) is given by

$$\dot{x} = \underbrace{\begin{pmatrix} -a_1 & 1 & & 0 \\ -a_2 & 0 & \ddots & \\ \vdots & \vdots & \ddots & 1 \\ -a_n & 0 & \cdots & 0 \end{pmatrix}}_{A_o} x + \underbrace{\begin{pmatrix} b_1 \\ b_2 \\ \vdots \\ b_n \end{pmatrix}}_{B_o} u; \quad y = \underbrace{(1 \;\; 0 \;\; \cdots \;\; 0)}_{C_o} x; \quad D_o = d \quad (2.27)$$

In this form, $A_o = A_c^T$; $B_o = C_c^T$; $C_o = B_c^T$.

Theorem 2.7
The system in observer canonical form is always observable, and its observability matrix satisfies

$$\mathcal{O}_o^{-1} = \begin{pmatrix} 1 & 0 & \cdots & 0 \\ a_1 & \ddots & \ddots & \vdots \\ \vdots & \ddots & \ddots & 0 \\ a_n & \cdots & a_1 & 1 \end{pmatrix} \tag{2.28}$$

The system realization is observable if and only if the realization can be transformed to an observer canonical form. The appropriate transformation is

$$T = \mathcal{O}^{-1}\mathcal{O}_o = (\mathcal{O}_o^{-1}\mathcal{O})^{-1} \tag{2.29}$$

If matrix A is diagonalizable, a *diagonal (modal) canonical form* for the transfer function $G(s) = d + \sum_{i=1}^{n} \frac{\beta_i \gamma_i}{s - s_i}$ (given in partial fraction expansion form [Appendix F.9]) is

$$\dot{x} = \underbrace{\begin{pmatrix} s_1 & 0 & \cdots & 0 \\ 0 & s_2 & \ddots & \vdots \\ \vdots & \ddots & \ddots & 0 \\ 0 & \cdots & 0 & s_n \end{pmatrix}}_{A_d} x + \underbrace{\begin{pmatrix} \beta_1 \\ \beta_2 \\ \vdots \\ \beta_n \end{pmatrix}}_{B_d} u; \quad y = \underbrace{(\gamma_1 \ \gamma_2 \ \cdots \ \gamma_n)}_{C_d} x; \quad D_d = d \tag{2.30}$$

NOTE The choice of β_i and γ_i could be arbitrary as far as their multiplication fits the numerator coefficients in the transfer function. It is convenient to choose all $\beta_i = 1$.

Theorem 2.8
1. The system is minimal if and only if $\beta_i \gamma_i \neq 0$ for all i.
2. The system is controllable if and only if $\forall i: \beta_i \neq 0$ and there is no multiplicity of eigenvalues.
3. The system is observable if and only if $\forall i: \gamma_i \neq 0$ and there is no multiplicity of eigenvalues.

When all eigenvalues of the system are different, diagonalization always exists. This is not the case when some eigenvalues are repeated. If the diagonalization does not exist, the *Jordan realizations* (near diagonal; with 1s above the main diagonal) are possible. For example, if the root s_1 is repeated 3 times, in this realization,

$$\dot{x} = \underbrace{\begin{pmatrix} s_1 & 1 & 0 & 0 & \cdots & 0 \\ 0 & s_1 & 1 & \vdots & & \vdots \\ \vdots & \ddots & s_1 & 0 & & \vdots \\ & & \ddots & s_4 & \ddots & \vdots \\ \vdots & & & \ddots & \ddots & 0 \\ 0 & \cdots & \cdots & \cdots & 0 & s_n \end{pmatrix}}_{A_J} x + \underbrace{\begin{pmatrix} 0 \\ 0 \\ 1 \\ \vdots \\ 1 \end{pmatrix}}_{B_J} u; \quad y = \underbrace{(\gamma_1 \ \gamma_2 \ \cdots \ \gamma_n)}_{C_J} x; \quad D_J = d \tag{2.31}$$

Chapter Two

Question
What is the controllability subspace of a controllable second-order system?

Answer
Since the second-order controllable system can have any x_1 and x_2, the controllability subspace will be the entire 2D space ($\mathbb{R}^2$).

Question
How would canonical state realizations change for a discrete-time system?

Answer
There is no significant change. The variable s will be replaced with z, and block diagrams will have delays instead of integrators.

Solved Problems

Problem 2.1
The discrete-time system realization is given by

$$x[k+1] = Ax[k] + Bu[k]$$
$$y[k] = Cx[k]$$

where $A = \begin{pmatrix} 0.1 & 0.1 \\ 0.1 & 0.1 \end{pmatrix}$; $B = \begin{pmatrix} 1 \\ 0 \end{pmatrix}$; $C = (1 \quad 0)$.

A. Check the stability of the system.
B. Check the controllability and observability of the system. Is the system minimal?
C. Find the transformation to the canonical controller form (if possible).
D. Find the transformation to the canonical observer form (if possible).
E. Find the realization of the canonical controller, observer, and diagonal forms for the given system.
F. How is it possible to move from the initial state $x_0 = (1,1)^T$ to the final state $x_f = (5,5)^T$?

Solution

A. The eigenvalues of the matrix $\begin{pmatrix} 0.1 & 0.1 \\ 0.1 & 0.1 \end{pmatrix}$ are computed from the characteristic polynomial equation $\det(zI - A) = z^2 - tr(A)z + \det(A) = z^2 - 0.2z + 0 = 0$. The roots of this quadratic equation are the eigenvalues $z_1 = 0$ and $z_2 = 0.2$. Both are inside the unit circle; therefore, the system is stable both asymptotically and BIBO.

B. The controllability matrix is $\mathcal{C} = \begin{pmatrix} 1 & 0.1 \\ 0 & 0.1 \end{pmatrix}$. The matrix is full rank (rank($\mathcal{C}$) = 2) which could be seen directly from this matrix. If the matrix is invertible, it must have full rank. A few alternative options for checking matrix invertibility are given in Appendix G.14. Here we just note that this matrix is upper triangular with no zeros on its main diagonal, and thus invertible. Consequently, the system is controllable.

The observability matrix is $\mathcal{O} = \begin{pmatrix} 1 & 0 \\ 0.1 & 0.1 \end{pmatrix}$ and it is full rank, because it is lower triangular with no zeros on its main diagonal, or because two rows are not proportional (linearly independent). Thus, the system is observable. Based on Theorem 2.5, controllable and observable systems must be minimal.

C. The transform of a controllable system to a canonical controller form is given by $T = \mathcal{C}\mathcal{C}_c^{-1}$ in (2.26).

$$T = \mathcal{C}\mathcal{C}_c^{-1} = \begin{pmatrix} 1 & 0.1 \\ 0 & 0.1 \end{pmatrix} \underbrace{\begin{pmatrix} 1 & -0.2 \\ 0 & 1 \end{pmatrix}}_{(2.25)} = \begin{pmatrix} 1 & -0.1 \\ 0 & 0.1 \end{pmatrix}$$

We will denote this transform matrix by T_c.

D. The transformation of an observable system to a canonical observer form is given by $T = (\mathcal{O}_o^{-1}\mathcal{O})^{-1}$ in (2.29).

$$T = (\mathcal{O}_o^{-1}\mathcal{O})^{-1} = \left(\underbrace{\begin{pmatrix} 1 & 0 \\ -0.2 & 1 \end{pmatrix}}_{(2.28)} \begin{pmatrix} 1 & 0 \\ 0.1 & 0.1 \end{pmatrix} \right)^{-1} = \begin{pmatrix} 1 & 0 \\ 1 & 10 \end{pmatrix}$$

We will denote this transformation matrix by T_o.

E. To find the controller and observer canonical realizations, we could use (2.22):

$$A_c = T_c^{-1}AT_c = \begin{pmatrix} 0.2 & 0 \\ 1 & 0 \end{pmatrix}; \; B_c = T_c^{-1}B = \begin{pmatrix} 1 \\ 0 \end{pmatrix}; \; C_c = CT_c = (1 \; -0.1); \; D_c = 0$$

$$A_o = T_o^{-1}AT_o = \begin{pmatrix} 0.2 & 1 \\ 0 & 0 \end{pmatrix} = A_c^T; \; B_o = T_o^{-1}B = \begin{pmatrix} 1 \\ -0.1 \end{pmatrix} = C_c^T; \; C_o = CT_o = (1 \; 0) = B_c^T; \; D_o = 0$$

Note that we could alternatively write these matrices directly from the transfer function computation (2.14) and templates in (2.24) and (2.27). The transfer function is

$$G(z) = C(zI - A)^{-1}B + D = (1 \; 0)\left(z\begin{pmatrix} 1 & 0 \\ 0 & 1 \end{pmatrix} - \begin{pmatrix} 0.1 & 0.1 \\ 0.1 & 0.1 \end{pmatrix} \right)^{-1}\begin{pmatrix} 1 \\ 0 \end{pmatrix} + 0 = \frac{z - 0.1}{z^2 - 0.2z + 0}$$

Thus, the numerator coefficients are $b_1 = 1$ and $b_2 = -0.1$ and the denominator coefficients are $a_1 = -0.2$ and $a_2 = 0$. If we substitute those numbers into (2.24) and (2.27), we get the same results as above.

For a diagonal canonical realization, we need to find partial fraction expansion of the transfer function:

$$G(z) = \frac{z - 0.1}{z^2 - 0.2z} = \frac{0.5}{z - 0} + \frac{0.5}{z - 0.2}$$

Now, we arbitrarily choose the coefficients β and γ in such a way that the appropriate multiplication $\beta_i \gamma_i$ will be 0.5 for $i = 1, 2$, as follows from the partial fraction expansion. For example, using (2.30):

$$A_d = \begin{pmatrix} 0 & 0 \\ 0 & 0.2 \end{pmatrix}; \; B_d = \begin{pmatrix} 1 \\ 1 \end{pmatrix}; \; C_d = (0.5 \; 0.5); \; D_d = 0$$

F. To get from state $x[0] = x_0 = \begin{pmatrix} 1 \\ 1 \end{pmatrix}$ to $x_f = \begin{pmatrix} 5 \\ 5 \end{pmatrix}$, the following input is required based on (2.20):

$$C^{-1}(x_f - A^2 x_0) = \begin{pmatrix} u[1] \\ u[0] \end{pmatrix} = \begin{pmatrix} 1 & 0.1 \\ 0 & 0.1 \end{pmatrix}^{-1} \left[\begin{pmatrix} 5 \\ 5 \end{pmatrix} - \begin{pmatrix} 0.1 & 0.1 \\ 0.1 & 0.1 \end{pmatrix}^2 \begin{pmatrix} 1 \\ 1 \end{pmatrix} \right] = \begin{pmatrix} 0 \\ 49.6 \end{pmatrix}$$

This means that first we need to input 49.6 to the system and then 0.

An interesting question here is whether or not the system would be at a desired state right after the first input (since the second input is zero). The answer is "no"; the second input is required to get to the desired state. Let's show that directly by iterating through the system's states:

$$x[1] = \begin{pmatrix} 0.1 & 0.1 \\ 0.1 & 0.1 \end{pmatrix} x[0] + \begin{pmatrix} 1 \\ 0 \end{pmatrix} u[0] = \begin{pmatrix} 0.1 & 0.1 \\ 0.1 & 0.1 \end{pmatrix} \begin{pmatrix} 1 \\ 1 \end{pmatrix} + \begin{pmatrix} 1 \\ 0 \end{pmatrix} 49.6 = \begin{pmatrix} 49.8 \\ 0.2 \end{pmatrix}$$

Obviously, this is not the desired state. Now, let's use the second input:

$$x[2] = \begin{pmatrix} 0.1 & 0.1 \\ 0.1 & 0.1 \end{pmatrix} x[1] + \begin{pmatrix} 1 \\ 0 \end{pmatrix} u[1] = \begin{pmatrix} 0.1 & 0.1 \\ 0.1 & 0.1 \end{pmatrix} \begin{pmatrix} 49.8 \\ 0.2 \end{pmatrix} + \begin{pmatrix} 1 \\ 0 \end{pmatrix} 0 = \begin{pmatrix} 5 \\ 5 \end{pmatrix}$$

and the system reaches the desired state.

Problem 2.2

The continuous state-space system is given by

$$A = \begin{pmatrix} -7 & 1 \\ -10 & 0 \end{pmatrix} \quad B = \begin{pmatrix} 3 \\ 15 \end{pmatrix} \quad C = (1 \ 0) \quad D = 0$$

Which of the following systems are similar (has similarity transform)? If the system is similar, then find the similarity transform T, and if not, then explain why.

A. $A = \begin{pmatrix} -7 & -10 \\ 1 & 0 \end{pmatrix} \quad B = \begin{pmatrix} 1 \\ 0 \end{pmatrix} \quad C = (3 \ 15) \quad D = 0$

B. $A = \begin{pmatrix} 0 & -10 \\ 1 & -7 \end{pmatrix} \quad B = \begin{pmatrix} 15 \\ 3 \end{pmatrix} \quad C = (0 \ 1) \quad D = 0$

C. $A = \begin{pmatrix} -2 & 0 \\ 0 & -5 \end{pmatrix} \quad B = \begin{pmatrix} 1 \\ 3 \end{pmatrix} \quad C = (3 \ 0) \quad D = 0$

D. $A = \begin{pmatrix} -2 & 0 \\ 0 & -5 \end{pmatrix} \quad B = \begin{pmatrix} 1 \\ 0 \end{pmatrix} \quad C = (2 \ 3) \quad D = 0$

Solution

The original system is given in a canonical observer form; thus, it is observable. The transfer function is $G(s) = C(sI - A)^{-1} B + D = \dfrac{3s + 15}{s^2 + 7s + 10} = \dfrac{3(s+5)}{(s+2)(s+5)} = \dfrac{3}{s+2}$. The system is not minimal and observable; thus, it is not controllable (Theorem 2.5).

A. The same system is given in a canonical controller form; it is controllable and not observable. Thus, the systems are not similar.

B. Pay attention that this system is exactly like the original system with states redefined in the opposite order, that is, x_1 is defined as x_2 and x_2 is defined as x_1. You could find such definition of canonical forms in some textbooks where the order of state variables is the opposite of what we have defined in this book. Nevertheless, the canonical forms defined that way still have the same properties. So, the new state $\tilde{x}$ is defined as $\tilde{x} = T^{-1}x = \begin{pmatrix} x_2 \\ x_1 \end{pmatrix}$ and the appropriate transform is $T = T^{-1} = \begin{pmatrix} 0 & 1 \\ 1 & 0 \end{pmatrix}$. The systems are similar.

C. The system is given in a diagonal canonical form with nonzero B elements; thus, the system is controllable, and not similar to the original system.

D. The transfer function is $G(s) = \dfrac{2s+10}{s^2+7s+10}$, which is different from the original system; thus, the systems are not similar.

Problem 2.3

The discrete system is given by

$$x[k+1] = Ax[k] + Bu[k]$$
$$y[k] = Cx[k]$$

$$A = \begin{pmatrix} 1 & 0 & 0 \\ a & 2 & 0 \\ b & 1 & 5 \end{pmatrix}; \quad B = \begin{pmatrix} 1 \\ 0 \\ 0 \end{pmatrix}; \quad C = (0 \ 1 \ 0)$$

A. Compute the transfer function $G(z)$ as a function of a and b variables.

B. What are the conditions on a and b such that the system will be controllable?

C. What are the conditions on a and b such that the system will be observable?

D. When the system is *not* observable, find two different initial conditions for which the outputs will be indiscernible given the input.
   For the rest of this question assume $a = 1$ and $b = 0$.

E. Let's assume that $x[0] = (1 \ 0 \ 0)^T$. Is there any input $u[0], u[1], u[2]$ that gives $x[3] = (0 \ 0 \ 0)^T$? If yes, then find that input sequence, and if not, then explain why not.

F. Repeat part (E) for $x[15] = (0 \ 0 \ 1)^T$, and find the appropriate input if possible.

Solution

A. Using Formula (2.14), the transfer function is

$$G(z) = C(zI - A)^{-1}B = (0 \ 1 \ 0) \begin{pmatrix} z-1 & 0 & 0 \\ -a & z-2 & 0 \\ -b & -1 & z-5 \end{pmatrix}^{-1} \begin{pmatrix} 1 \\ 0 \\ 0 \end{pmatrix}$$

$$= \dfrac{a(z-5)}{(z-1)(z-2)(z-5)}$$

B. The system is controllable if the determinant of a controllability matrix is not zero.

$$\mathcal{C} = \begin{pmatrix} 1 & 1 & 1 \\ 0 & a & 3a \\ 0 & b & a+6b \end{pmatrix}$$

$$\det(\mathcal{C}) = a(a+6b) - 3ab = a^2 + 6ab - 3ab = a^2 + 3ab = a(a+3b) \neq 0$$

The required condition is

$$\{a \neq 0\} \text{ and } \{a \neq -3b\}$$

C. The observability matrix is

$$\mathcal{O} = \begin{pmatrix} 0 & 1 & 0 \\ a & 2 & 0 \\ 3a & 4 & 0 \end{pmatrix} \Rightarrow \det(\mathcal{O}) = 0 \quad \text{(The system is not observable.)}$$

D. It can be seen directly from the equations that x_3 is not affecting any other state variable or the output. Thus, any initial conditions with different $x_3(0)$ values will be indiscernible from the output measurement.

E. Using (2.20), we can get to zero state in three steps by providing the following inputs:

$$u = \mathcal{C}^{-1}(x[3] - A^3 x[0])$$

where $\mathcal{C} = [B \quad AB \quad A^2 B] = \begin{pmatrix} 0 & 1 & 0 \\ 1 & 2 & 0 \\ 3 & 4 & 0 \end{pmatrix} \Rightarrow \mathcal{C}^{-1} = \begin{pmatrix} 1 & -1 & 2 \\ 0 & 1 & -3 \\ 0 & 0 & 1 \end{pmatrix}$

$$u[0:2] = (-8 \quad 17 \quad -10)^T$$

F. We know that the system needs at the most three steps to get to any state; thus, we need three steps to get to zero (see [E] above), and then nine steps which will keep "0" state, and, finally, three more steps to get the final state:

$$u = \mathcal{C}^{-1}(x[15] - A^3 x[12]) = \mathcal{C}^{-1} x[15] = \begin{pmatrix} 1 & -1 & 2 \\ 0 & 1 & -3 \\ 0 & 0 & 1 \end{pmatrix} \begin{pmatrix} 0 \\ 0 \\ 1 \end{pmatrix} = \begin{pmatrix} 2 \\ -3 \\ 1 \end{pmatrix}$$

$$u[0:14] = (-8 \quad 17 \quad -10 \quad 0 \quad 0 \quad 0 \quad 0 \quad 0 \quad 0 \quad 0 \quad 0 \quad 1 \quad -3 \quad 2)^T$$

CHAPTER 3
Pole Placement via State Feedback

Classical control tools allow us to design a serial or a feedback controller that stabilizes the system and allows reaching the desired performance characteristics. The tricky part here is that this basic design rather resembles black art where a lot of trial and error are needed to get anywhere close to what we want. There is no easy way to design a controller using the classical theory even with modern computational power. In turn, modern control allows designing a controller using a single formula or one line of code. The catch here is that it is not a serial or a feedback controller, but a state-space controller which has a different architecture. In this chapter, we will discuss feedback and three different ways to design a controller. The main goal here is to design a controller that will place all closed-loop poles of the system at the desired locations, which will in turn provide both stability and performance characteristics.

State Feedback

Given the system

$$\begin{cases} \dot{x} = Ax + Bu \\ y = Cx + Du \end{cases} \quad (3.1)$$

we would like to design a state feedback controller K that will satisfy all the requirements of stability and performance.

NOTE We could use two equivalent block diagrams for the open-loop system representation as shown in Figure 3.1. For simplicity, we will frequently assume that $D = 0$ and that the blocks related to D will disappear.

The idea is to return via the feedback the weighted sum of the states x. Thus, the controller $K_{1 \times n}$ is a row vector of constant numbers multiplied by a vector x (inner product) as shown in Figure 3.2:

$$u(t) = r(t) - K \cdot x(t) \quad (3.2)$$

CAUTION! It is important to remember that the vector of the controller's gains K is defined as a row (and not a column) vector.

36   Chapter Three

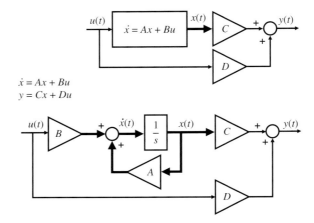

$\dot{x} = Ax + Bu$
$y = Cx + Du$

FIGURE 3.1 Two equivalent representations of the open-loop state-space system given in (3.1). (Bold lines denote vectors.)

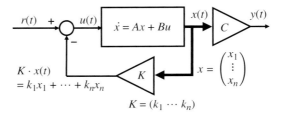

FIGURE 3.2 State-space feedback closed-loop control architecture.

Substituting the definition of input u from (3.2) into (3.1),

$$\begin{cases} \dot{x} = Ax + B(r - Kx) = (A - BK)x + Br \\ y = Cx + D(r - Kx) = (C - DK)x + Dr \end{cases} \quad (3.3)$$

Thus, the closed-loop system from $r(t)$ to $y(t)$ is equivalent to the system with the input $r(t)$, and matrices $\{\tilde{A} = A - BK, \tilde{B} = B, \tilde{C} = C - DK, \tilde{D} = D\}$. These matrices represent a closed-loop system. We want to choose the values in the vector K in a way that the closed-loop poles of the system will be at desired locations. In other words, we want to choose the eigenvalues of the closed-loop matrix $\tilde{A} = A - BK$ using the appropriate vector K.

NOTE The closed-loop transfer function is

$$G_{cl}(s) = \frac{\text{open-loop zeros}}{\text{closed-loop poles}} = \frac{\det(sI - A + B(K + C - DK))}{\det(sI - A + BK)} + D - 1 \quad (3.4)$$

CAUTION! State feedback controller *cannot* change the system's zeros. The only way to change zeros is by cancelation of closed-loop poles and open-loop zeros.

Controller Design

Open-loop poles are the roots of A's characteristic polynomial $\det(sI - A)$. We denote this open-loop polynomial (denominator of the transfer function) by $a(s) = s^n + a_1 s^{n-1} + \cdots + a_n$, or by a vector of the coefficients $\vec{a} = (a_1, a_2, \ldots, a_n)$. Now, we want the poles in the closed loop to be at the predefined locations s_i. Thus, the desired closed-loop polynomial is $\alpha(s) = (s-s_1)(s-s_2)\cdots(s-s_n) = s^n + \alpha_1 s^{n-1} + \cdots + \alpha_n$. We denote the vector of its coefficients by $\vec{\alpha} = (\alpha_1, \alpha_2, \ldots, \alpha_n)$. The following theorems explain how to design a controller for a controllable system.

Theorem 3.1 (Bass-Gura, 1965)

$$K = (\vec{\alpha} - \vec{a}) \mathcal{C}_c \mathcal{C}^{-1} \tag{3.5}$$

where $\mathcal{C}$ is a controllability matrix of the system, and $\mathcal{C}_c$ is the canonical controller controllability matrix.

Theorem 3.2 (Ackermann, 1972)

$$K = (0 \quad 0 \quad \cdots \quad 0 \quad 1)\mathcal{C}^{-1}\alpha(A) \tag{3.6}$$

that is, the controller gain K is the last row of $\mathcal{C}^{-1}\alpha(A)$ matrix.

It is possible to design the controller gain K using one of the theorems above, but the simplest method for low-order systems is to solve the problem by comparing the coefficients of the desired closed-loop characteristic polynomials $\alpha(s)$ and the actual polynomial $\det(sI - (A - BK)) = \det(sI - A + BK)$.

Design Algorithm

1. Choose the appropriate location of the closed-loop poles $s_i \in \mathbb{C}$.
2. Compute the desired closed-loop polynomial:

$$\alpha(s) = (s-s_1)(s-s_2)\cdots(s-s_n) = s^n + \alpha_1 s^{n-1} + \cdots + \alpha_n$$

3. Method A: Compute the coefficients $a(s) = \det(sI - A)$ and $\alpha(s)$. Then, compute matrices $\mathcal{C}_c$ and $\mathcal{C}^{-1}$, and use the Bass-Gura formula.
4. Method B: Compute $\alpha(A)$ and $\mathcal{C}^{-1}$. Then apply Akermann's formula.
5. Method C: Compute $a_K(s) = \det(sI - A + BK) = s^n + a_1(k_1, \ldots, k_n)s^{n-1} + \cdots + a_n(k_1, \ldots, k_n)$ and compare the coefficients to the polynomial $\alpha(s)$. Solve the obtained system of equations $a_i(k_1, \ldots, k_n) = \alpha_i$ to extract the $K = (k_1, \ldots, k_n)$.

NOTE For a controllable system we are able to place the poles wherever we want.

NOTE It is not readily clear how to choose the locations of the closed-loop poles. In general, it is advisable to choose the locations by dominant pole analysis, or at the locations of the Bessel filter (on the left semicircle in the complex domain).

VERY IMPORTANT NOTE For discrete systems, almost all the learned methods and formulas are the same, except for the change of s to z, the change of $\dot{x}(t)$ to $x(k+1)$, and the unit circle stability region instead of the left semi-plane. Moreover, the same methods of design can be applied to multiple-input multiple-output (MIMO) systems; in that case the gain K is a matrix, and there are some additional degrees of freedom in choosing the gains.

Tracking the Input Signal

CAUTION! After learning control systems for a while, it might become second nature to assume that the system's output in a closed loop with a stabilizing controller closely follows the reference input. In other words, for a unit step input we expect to see the output converging eventually to 1. Obviously, it is a misconception to think that all stable closed-loop systems are like that. The steady-state error could be arbitrarily big in the closed loop (depends on the controller and plant).

Suppose we have a unit step input [reference signal $r(t)$], and we want to compute the steady-state error. Of course, we want this error to be as small as possible for a good tracking. Using the relations (3.3), we can compute the closed-loop transfer function $G_{cl}(s) = \tilde{C}(sI - \tilde{A})^{-1}\tilde{B} + D = (C - DK)(sI - A + BK)^{-1}B + D$. Thus, the DC gain (for $s = 0$) is

$$G_{cl}(0) = (C - DK)(BK - A)^{-1}B + D \tag{3.7}$$

which is definitely not 1 in most cases.

Question
If we want zero steady-state error, that is, the DC gain of 1 in the closed loop, can we multiply the reference signal r by the reciprocal of $G_{cl}(0)$ in (3.7) to make it happen?

Solution
Yes, this trivial solution will work, but there are serious downsides to this solution. First, the gain $1/G_{cl}(0)$ will multiply noise and disturbances outside the loop. Second, we cannot know the model precisely (or the physical model could change in time) and the actual DC gain could be different from the modeled DC gain, so $1/G_{cl}(0)$ gain will not cancel the actual steady-state error precisely.

Integrator in the Loop

If we want to ensure zero steady-state error for a step response, we have to add an integrator in the loop as shown in Figure 3.3.

We need to rewrite the system's equations to make it possible to design a state feedback controller with the additional state x_{n+1}. Clearly, the system is of the order $n+1$ now.

If we define the new state as

$$\bar{\bar{x}} = \begin{pmatrix} x \\ x_{n+1} \end{pmatrix} \tag{3.8}$$

Pole Placement via State Feedback

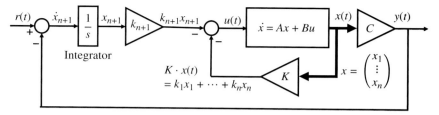

FIGURE 3.3 Integrator in the loop-controller design for zero steady-state error.

we obtain

$$\bar{\bar{x}} = \begin{pmatrix} \dot{x} \\ \dot{x}_{n+1} \end{pmatrix} = \begin{pmatrix} Ax + Bu \\ r - Cx \end{pmatrix} = \begin{pmatrix} A & 0 \\ -C & 0 \end{pmatrix} \begin{pmatrix} x \\ x_{n+1} \end{pmatrix} + \begin{pmatrix} B \\ 0 \end{pmatrix} u + \begin{pmatrix} 0 \\ 1 \end{pmatrix} r \quad (3.9)$$

Thus, $\bar{\bar{A}} = \begin{pmatrix} A & 0 \\ -C & 0 \end{pmatrix}$, and $\bar{\bar{B}} = \begin{pmatrix} B \\ 0 \end{pmatrix}$. We can design a controller of the order $n+1$ for these matrices $\bar{\bar{K}} = (K, k_{n+1})$ using the usual tools described previously in this chapter.

Question
Is it possible to stabilize a system if that system is not controllable?

Answer
Yes, it is possible to stabilize it, but not always. Some states might not be controllable but converge naturally by themselves. If this is the case, the system could be stabilized. Refer to Chapter 5 for more information about such systems.

Question
What are the disadvantages of the state feedback controller design compared to the classical control methods (Bode diagrams, root locus, etc.)?

Answer
The state-space design requires availability of all the system's states. For most physical systems, this would require multiple sensors and/or computational modules. We will explain how to relax this requirement in Chapter 4.

Question
Pay attention that we ignore matrix C in our controller design. Does this mean that we can change C to another matrix (different transfer function) and still get the same response specifications?

Answer
This question is tricky. Yes, we can change matrix C to any other matrix without changing the location of closed-loop poles. The thing is that matrix C affects the location of open-loop zeros of the system and those are the same zeros in the closed loop since state feedback does not change the location of zeros. We know from the classical control theory that zeros affect the response, though less than dominant poles. Thus, we will not get the same response specifications.

Solved Problems

Problem 3.1
Let's return to Problem 2.1 from Chapter 2 and design a stabilizing state controller. The discrete system realization is given by

$$x[k+1] = Ax[k] + Bu[k]$$
$$y[k] = Cx[k]$$

$$A = \begin{pmatrix} 0.1 & 0.1 \\ 0.1 & 0.1 \end{pmatrix}; \quad B = \begin{pmatrix} 1 \\ 0 \end{pmatrix}; \quad C = (1 \quad 0)$$

Use three different methods to design a controller where all the closed-loop poles will be at −0.1.

Solution
The first method uses the Bass-Gura formula. To use that formula, we need to identify the open-loop characteristic polynomial and its coefficients, the desired closed-loop characteristic polynomial of the same degree and its coefficients, and the controllability matrix.

The controllability matrix is given by $\mathcal{C} = \begin{pmatrix} 1 & 0.1 \\ 0 & 0.1 \end{pmatrix}$. The open-loop characteristic polynomial is $a(z) = z^2 - 0.2z$, and to locate both closed-loop poles at −0.2 the desired polynomial is $\alpha(z) = (z+0.1)^2 = z^2 + 0.2z + 0.01$. Consequently, the coefficient vectors are $\vec{a} = (-0.2 \quad 0)$ and $\vec{\alpha} = (0.2 \quad 0.01)$. Using the Bass-Gura formula (3.5),

$$K = (\vec{\alpha} - \vec{a})\mathcal{C}_c \mathcal{C}^{-1} = ((0.2 \quad 0.01) - (-0.2 \quad 0)) \begin{pmatrix} 1 & 0.2 \\ 0 & 1 \end{pmatrix} \begin{pmatrix} 1 & 0.1 \\ 0 & 0.1 \end{pmatrix}^{-1} = (0.4 \quad 0.5)$$

The controller gains are $k_1 = 0.4$ and $k_2 = 0.5$. Note that the controllability matrix and canonical controller controllability matrix were computed earlier in Problem 2.1.

Now, we will try to get the same result using the Ackermann formula. Given that $\alpha(z) = z^2 + 0.4z + 0.04$ as before, the $\alpha(A)$ would be defined as $A^2 + 0.2A + 0.01I$. Based on (3.6), we need to find the last row of the matrix $\mathcal{C}^{-1}\alpha(A) = \begin{pmatrix} 1 & 0.1 \\ 0 & 0.1 \end{pmatrix}^{-1} \left[\begin{pmatrix} 0.1 & 0.1 \\ 0.1 & 0.1 \end{pmatrix}^2 + \right.$
$\left. 0.2 \begin{pmatrix} 0.1 & 0.1 \\ 0.1 & 0.1 \end{pmatrix} + 0.01 \begin{pmatrix} 1 & 0 \\ 0 & 1 \end{pmatrix} \right] = \begin{pmatrix} 0.01 & -0.01 \\ 0.4 & 0.5 \end{pmatrix}$, which is the same vector $K = (0.4 \quad 0.5)$.

The third way to compute the controller gains is by a comparison of coefficients of the desired closed-loop polynomial $\alpha(z)$ and the actual closed-loop characteristic polynomial $\det(zI - A + BK)$.

$$\det(zI - A + BK) = \det\left(\begin{pmatrix} z & 0 \\ 0 & z \end{pmatrix} - \begin{pmatrix} 0.1 & 0.1 \\ 0.1 & 0.1 \end{pmatrix} + \begin{pmatrix} 1 \\ 0 \end{pmatrix}(k_1 \quad k_2) \right)$$

$$= \det\begin{pmatrix} z - 0.1 + k_1 & -0.1 + k_2 \\ -0.1 & z - 0.1 \end{pmatrix} = (z - 0.1 + k_1)(z - 0.1) + 0.1(k_2 - 0.1)$$

$$= z^2 + (k_1 - 0.2)z + (-0.1k_1 + 0.1k_2)$$

where $K = (k_1 \quad k_2)$.

Now, we can compare coefficients with $\alpha(z) = z^2 + 0.2z + 0.01$:

$$\begin{cases} 0.2 = k_1 - 0.2 \\ 0.01 = 0.1(k_2 - k_1) \end{cases}$$

From the first equation, $k_1 = 0.4$, and from the second equation, $k_2 = 0.5$, as expected.

Problem 3.2
The system is defined by

$$\dot{x}(t) = Ax(t) + Bu(t)$$
$$y(t) = Cx(t) + Du(t)$$

$$A = \begin{pmatrix} -1 & -1 & -1 \\ 1 & 0 & 0 \\ 0 & 1 & 1 \end{pmatrix} \quad B = \begin{pmatrix} 1 \\ 0 \\ 0 \end{pmatrix} \quad C = (3 \ 7 \ 9) \quad D = 1$$

A. Is this system controllable? Is it observable?

B. Assume that the entire state is measurable. We close the loop with the feedback controller: $u(t) = r(t) - Kx(t)$. Design the state feedback controller K such that all the closed-loop poles will be at -1.

C. Compute the closed-loop transfer function of the system.

D. What is the simplest correction to the system that is needed to get a zero steady-state error with step response?

Solution

A. The controllability and observability matrices are

$$\mathcal{C} = \begin{pmatrix} 1 & -1 & 0 \\ 0 & 1 & -1 \\ 0 & 0 & 1 \end{pmatrix} \quad \Rightarrow \quad \text{rank}(\mathcal{C}) = 3$$

$$\mathcal{O} = \begin{pmatrix} 3 & 7 & 9 \\ 4 & 6 & 6 \\ 2 & 2 & 2 \end{pmatrix} \quad \Rightarrow \quad \text{rank}(\mathcal{O}) = 3$$

The system is controllable and observable.

B. The desired closed-loop characteristic polynomial is

$$\alpha(s) = s^3 + 3s^2 + 3s + 1$$

$$a_k(s) = \det(sI - A + BK) = \det \begin{pmatrix} s + k_1 + 1 & k_2 + 1 & k_3 + 1 \\ -1 & s & 0 \\ 0 & -1 & s - 1 \end{pmatrix}$$

$$= (s + k_1 + 1)s(s - 1) + ((s - 1)(k_2 + 1) + k_3 + 1) = s^3 + k_1 s^2 + (k_2 - k_1)s + (k_3 - k_2)$$

From the comparison of the polynomial coefficients, $k_1 = 3$, $k_2 = 6$, $k_3 = 7$.

C. Closed-loop matrices are $A_{cl} = A - BK, B_{cl} = B, C_{cl} = C - DK, D_{cl} = D$. Then, the transfer function is

$$G(s) = C_{cl}(sI - A_{cl})^{-1}B_{cl} + D_{cl} = \frac{s^2 + 2s + 2}{s^2 + 2s + 1} \cdot \frac{s+1}{s+1}$$

D. We can see from the transfer function in part (C) above that the DC gain (when $s = j\omega = 0$) is 2; thus, all the input step responses will be multiplied by 2 and the steady-state error for unit step will be 100 percent. To correct this problem, we need to multiply the reference signal r by ½.

Problem 3.3
Prove the Bass-Gura theorem.

Solution
We want to show that the eigenvalues of the closed-loop matrix $A - BK$ could be repositioned to the locations defined by the desired polynomial $\alpha(s)$ roots when using the controller $K = (k_1 \quad \cdots \quad k_n) = (\vec{\alpha} - \vec{a})\mathcal{C}_c \mathcal{C}^{-1}$.

First, we will prove it for a simple case where the system implementation is given in a controllable canonical form:

$$\dot{x} = \underbrace{\begin{pmatrix} -a_1 & -a_2 & \cdots & -a_n \\ 1 & 0 & \cdots & 0 \\ & \ddots & & \vdots \\ 0 & & 1 & 0 \end{pmatrix}}_{A_c} x + \underbrace{\begin{pmatrix} 1 \\ 0 \\ \vdots \\ 0 \end{pmatrix}}_{B_c} u$$

Here,

$$A_c - B_c K_c = \begin{pmatrix} -a_1 & -a_2 & \cdots & -a_n \\ 1 & 0 & \cdots & 0 \\ & \ddots & & \vdots \\ 0 & & 1 & 0 \end{pmatrix} - \begin{pmatrix} k_{c1} & k_{c2} & \cdots & k_{cn} \\ 0 & & \cdots & 0 \\ \vdots & \ddots & & \vdots \\ 0 & \cdots & & 0 \end{pmatrix}$$

$$= \begin{pmatrix} -a_1 - k_{c1} & -a_2 - k_{c2} & \cdots & -a_n - k_{cn} \\ 1 & 0 & \cdots & 0 \\ & \ddots & & \vdots \\ 0 & & 1 & 0 \end{pmatrix}$$

The characteristic polynomial of that matrix in the companion form is $\det(sI - A_c + B_c K_c) = s^n + (a_1 + k_{c1})s^{n-1} + (a_2 + k_{c2})s^{n-2} + \cdots + (a_n + k_{cn})$, and it should be equal to the desired polynomial $s^n + \alpha_1 s^{n-1} + \alpha_2 s^{n-2} + \cdots + \alpha_n$. From the comparison of the coefficients of two polynomials, we get $k_{ci} = \alpha_i - a_i; i = 1, 2, \ldots, n$, which is exactly the Bass-Gura formula where $\mathcal{C} = \mathcal{C}_{c'}$ thus $K_c = (\vec{\alpha} - \vec{a})\mathcal{C}_c \mathcal{C}^{-1} = (\vec{\alpha} - \vec{a})$.

Now, for a general controllable system $\{A, B, C, D\}$, we can always find the transformation to a controllable canonical form. Formula (2.26) gives that transformation as $T = \mathcal{C}\mathcal{C}_c^{-1}$.

We want to find such a K that $A - BK$ will have eigenvalues at the locations of $\alpha(s)$ roots. The characteristic polynomial of $A - BK$ and $A_c - B_c K_c$ should be the same, since the similarity transform is not changing the eigenvalues. We can write the equality of characteristic polynomials as $\det(sI - A + BK) = \det(sI - A_c + B_c K_c)$. If we substitute the transformed matrices in the left determinant, we will get $\det(sI - A + BK) = \det(sI - TA_c T^{-1} + TB_c K) = \det(T^{-1})\det(sI - TA_c T^{-1} + TB_c K)\det(T) = \det(T^{-1}sIT - T^{-1}TA_c T^{-1}T + T^{-1}TB_c KT) = \det(sI - A_c + B_c KT) = \det(sI - A_c + B_c K_c)$. Thus, $KT = K_c$ and $K = K_c T^{-1} = K_c \mathcal{C}_c \mathcal{C}^{-1} = (\vec{\alpha} - \vec{a})\mathcal{C}_c \mathcal{C}^{-1}$, which concludes the proof.

CHAPTER 4
State Estimation (Observers)

So far, we have assumed that the full state x is available to us, that is, we can somehow measure all the variables included in the state vector x (e.g., by adding sensors). In practice, not all variables are available. Some quantities are hard to measure directly, while for others the sensors needed for the measurements are too expensive. Fortunately, in many cases we can estimate the state x based only on the input u and the output y. The mechanism (or algorithm) for such an estimation is called observer.

Observer Structure

It is important to remember that the plant is a physical (mechanical, electrical, chemical, etc.) system and that its outputs are measured by sensors. Though we draw a linear mathematical model instead of a plant, the actual output is not the output of the model, but the output of the real physical system. The observer is different—it is a simulation of a mathematical system's model that is trying to emulate the work of a real system. Since it is a simulated model, we have access to its states and hope that the real states are close enough. It would be nice if we could follow the states of the system based on a model.

Question
Suppose we know precisely the state-space model of a given system. In other words, matrices A, B, C, D are known. We plan to compute the states of the real system by simulating the system of differential equations $\dot{x} = Ax + Bu$ and using the same input u, as shown in Figure 4.1. Why would this approach for state estimation not work well?

Answer
The problem is that we don't know the system's initial conditions; therefore, even if we know the model precisely, this will not work. We start our observer's simulation from arbitrary initial conditions (generally zero initial conditions unless there is a good reason to choose something else). Since the physical plant and the simulated system start from different initial conditions, it is obvious that they will not converge to the same states.

An additional problem is that in practice, matrices A, B, C, D are known only approximately. So, the idea of observer is to start with zero initial conditions and try emulating the system's behavior by simulating its state equations, but at the same time comparing the emulated (estimated by model simulation) output to the real output and trying to minimize the difference between $y(t)$ and estimated $\hat{y}(t)$.

44 Chapter Four

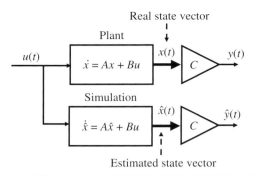

FIGURE 4.1 Incorrect way to estimate the state.

The observer equations proposed by David Luenberger in 1964 (Luenberger, 1964) are

$$\begin{cases} \dot{\hat{x}} = A\hat{x} + Bu + L(y - \hat{y}) = (A - LC)\hat{x} + Bu + Ly \\ \hat{y} = C\hat{x} + Du \end{cases} \quad (4.1)$$

where $\{A, B, C, D\}$ is the system's model, $\hat{x}$ denotes the estimated state, and $\hat{y}$ denotes the estimated output. The goal is to design a vector L such that the estimated state $\hat{x}$ will converge as fast as possible to the real state x of the system.

NOTE The observer's initial conditions are $\hat{x}(0) = 0$, unless otherwise explicitly stated.

Figure 4.2 shows a general structure of an observer implementing those equations. For simplicity, we assume $D = 0$. Make sure that you understand how different connections between the blocks implement the equations (compare to Figure 3.1).

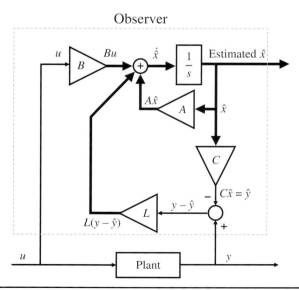

FIGURE 4.2 Linear observer estimating the states of a plant.

Question
Suppose our model is defined by matrices $\{A, B, C, D\}$. When does it make sense to have very high values in the vector L?

Answer
It makes sense when the measurement $y(t)$ is very reliable and the system's model not so much. It would give more weight to the output's error.

Observer Design

We want the observer's state $\hat{x}$ be as close as possible to the plant's state x as soon as possible. In other words, we want the estimation error $\tilde{x} = x(t) - \hat{x}(t)$ to be as small as possible (as fast as possible).

Theorem 4.1

$$\dot{\tilde{x}} = (A - LC)\tilde{x} \tag{4.2}$$

that is, the estimation error is fading according to the eigenvalues of the matrix $A - LC$.

Thus, we will want to design a vector L in such a way that the eigenvalues of $A - LC$ will be stable and "fast" (far to the left for the continuous-time systems or closer to origin for the discrete-time systems).

Question
Suppose we can choose the observer's eigenvalues of continuous-time system anywhere we want, and we want them as fast as possible. Is it wrong to choose them all in $-10^{1000000}$?

Answer
While theoretically it is possible to do so, practically it is a very bad idea. As always, using faster eigenvalues increases the system's bandwidth, which, in turn, makes our observer more sensitive to noise and unwanted disturbances.

NOTE For discrete-time systems the design process is the same (while replacing s with z).

NOTE Note the similarity between the problem of designing the eigenvalues of $A - BK$ (controller) and $A - LC$ (observer). Those are dual problems and similar techniques are used to solve them.

Theorem 4.2
If the system is observable, it is possible to design the observer with the vector of gains L, such that its poles are anywhere we want.

RULE OF THUMB It is advisable to choose the observer's poles (eigenvalues of $A - LC$) approximately 2 to 5 times to the left of the plant's closed-loop poles.

Observer's Design Algorithm
1. Choose the desired location of the observer's poles $\tilde{s}_i$ in the closed loop.
2. Compute the desired characteristic polynomial for observer in the closed loop:
$$\tilde{\alpha}(s) = (s - \tilde{s}_1)(s - \tilde{s}_2)\cdots(s - \tilde{s}_n) = s^n + \tilde{\alpha}_1 s^{n-1} + \cdots + \tilde{\alpha}_n$$
3. Method 1: Compute $a(s) = \det(sI - A) = s^n + a_1 s^{n-1} + \cdots + a_n$, $\mathcal{O}_o$ and $\mathcal{O}^{-1}$. Denote: $\tilde{\vec{\alpha}} = (\tilde{\alpha}_1, \ldots, \tilde{\alpha}_n)$; $\vec{a} = (a_1, \ldots, a_n)$ and use the following Bass-Gura formula:
$$L = \mathcal{O}^{-1}\mathcal{O}_o(\tilde{\vec{\alpha}} - \vec{a})^T \quad (4.3)$$
4. Method 2: Compute the vector of observer gains using the Ackermann formula:
$$L = \tilde{\alpha}(A)\mathcal{O}^{-1}\begin{pmatrix} 0 \\ \vdots \\ 0 \\ 1 \end{pmatrix} \quad (4.4)$$
that is, the last column of $\tilde{\alpha}(A)\mathcal{O}^{-1}$ matrix.
5. Method 3: Compare the coefficients of $\tilde{\alpha}(s)$ and $a_L(s) = \det(sI - A + LC) = s^n + a_1(l_1, \ldots, l_n)s^{n-1} + \cdots + a_n(l_1, \ldots, l_n)$ to obtain the system of linear equations. Solve this system with regard to the vector of observer gains $L = (l_1, \ldots, l_n)^T$.

Integrated System: State Feedback + Observer

Now, we are able to get all the advantages of an arbitrary pole placement using only the input and output of the open-loop plant. All that is left to do is to connect the estimated state (instead of the original state) to the controller gains K and return this weighted sum in the feedback loop. The block diagram of state feedback with the observer is shown in Figure 4.3 and with additional details in Figure 4.4.

Question
The real system and the observer start with different initial conditions. Will the system in closed loop with the state controller and observer be always stable if the initial conditions are very different?

Answer
Yes, the system will be stable with stabilizing controller regardless of the initial conditions. We know that a stable linear system converges globally from any initial conditions.

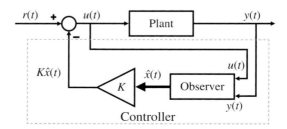

Figure 4.3 The block diagram of a closed-loop state feedback with the observer.

State Estimation (Observers)

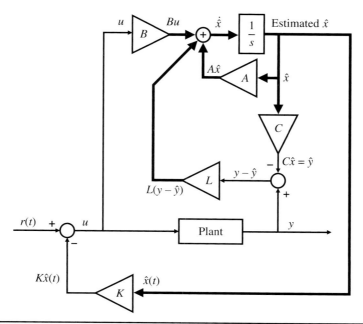

FIGURE 4.4 Detailed block diagram of a closed-loop state feedback with the observer.

Theorem 4.1 demonstrates that eventually the estimated state will converge to the real state. At that point of time, the state of the real system will be equal (or very close) to the estimated state. You can think about that time point as new "time zero." Since from that point on the estimated state will follow the real state (whatever it is), the system will converge as it would with a regular state feedback after some transient response.

Question
Notice that we design the controller as there is no observer (assuming that all states are accessible) and designing the observer as there is no controller and closed loop. Aren't controller and observer affecting each other? Why is it allowable to ignore their interrelations when designing them?

Answer
This is a tough question better answered by the following theorem. The magic thing about observers is that their structure is designed in such a way that observers do not change the closed-loop transfer function at all. It is difficult to believe, but we will demonstrate that soon.

Theorem 4.3 (Separation Principle)
There is no mutual dependence between the system's closed-loop poles and the observer's poles. Thus, the observer and controller can be designed independently.

Explanation

Using (3.1), (4.1), and the feedback definition $u = r - K\hat{x}$, the integrated system in Figure 4.4 is represented by the following equations:

$$\begin{cases} \dot{x} = Ax + Bu \\ \dot{\hat{x}} = (A - LC)\hat{x} + Bu + Ly = (A - LC)\hat{x} + Bu + LCx \\ u = r - K\hat{x} \end{cases}$$

After substitution of u into the first two equations

$$\begin{cases} \dot{x} = Ax + B(r - K\hat{x}) \\ \dot{\hat{x}} = (A - LC)\hat{x} + B(r - K\hat{x}) + LCx \end{cases}$$

or in matrix form

$$\begin{pmatrix} \dot{x} \\ \dot{\hat{x}} \end{pmatrix} = \begin{pmatrix} A & -BK \\ LC & A - LC - BK \end{pmatrix} \begin{pmatrix} x \\ \hat{x} \end{pmatrix} + \begin{pmatrix} B \\ B \end{pmatrix} r \qquad (4.5)$$

$$y = Cx = (C \quad 0) \begin{pmatrix} x \\ \hat{x} \end{pmatrix} \qquad (4.6)$$

Now let's apply the similarity transformation (not changing eigenvalues and transfer function):

$$\begin{pmatrix} x \\ \hat{x} \end{pmatrix} = \begin{pmatrix} I & 0 \\ I & -I \end{pmatrix} \begin{pmatrix} x \\ \tilde{x} \end{pmatrix} \qquad (4.7)$$

where I is the identity matrix, and get

$$\begin{pmatrix} \dot{x} \\ \dot{\tilde{x}} \end{pmatrix} = \begin{pmatrix} A - BK & BK \\ 0 & A - LC \end{pmatrix} \begin{pmatrix} x \\ \tilde{x} \end{pmatrix} + \begin{pmatrix} B \\ 0 \end{pmatrix} r \qquad (4.8)$$

$$y = (C \quad 0) \begin{pmatrix} x \\ \tilde{x} \end{pmatrix} \qquad (4.9)$$

Note that now the closed-loop matrix of the system with the controller and observer has upper block-triangular form; thus the estimation error is independent from the controller.

Theorem 4.4

The closed-loop transfer function (for $D = 0$) is

$$T(s) = \frac{Y(s)}{R(s)} = C(sI - A + BK)^{-1} B \qquad (4.10)$$

which is the same as the closed-loop transfer function without the observer.

Explanation

$$T(s) = \tilde{C}(sI - \tilde{A} + \tilde{B}K)^{-1} \tilde{B}$$

where tilde denotes the closed-loop block matrices defined in (4.8) and (4.9).

Thus,

$$T(s) = (C \quad 0) \begin{pmatrix} sI - A + BK & -BK \\ 0 & sI - A + LC \end{pmatrix}^{-1} \begin{pmatrix} B \\ 0 \end{pmatrix}$$

$$= (C \quad 0) \begin{pmatrix} (sI - A + BK)^{-1} & \text{something...} \\ 0 & (sI - A + LC)^{-1} \end{pmatrix} \begin{pmatrix} B \\ 0 \end{pmatrix} = C(sI - A + BK)^{-1} B$$

NOTES

1. The observer's poles are all canceled in the closed-loop transfer function.
2. The initial conditions for the observer are always chosen as zero $\hat{x}(0) = 0$, if not otherwise stated.
3. There are two sources of uncertainty in the system: the uncertainty in model parameters and noisy measurements. If the measurements are not too noisy, but the system's parameters have high uncertainty, then we would trust the output measurement more, and increase the weight of the $y - \hat{y}$ error (the observer's gain L), choosing the "faster" observer's poles.

Question
Think of an example of a physical system for which not all state variables can be measured.

Answer
There is an infinite number of potential examples from the velocity of a paper airplane to a turbine's airflow per second. In the first case, sensors would unnecessarily increase the weight and change the behavior of a small paper airplane. In the second case, sensors would be too expensive or unreliable.

Question
What will happen to a system in a closed loop with the observer's poles "slower" than the system's poles?

Answer
It will take longer time to estimate the correct state; thus, the transient system's response will be longer.

Question
Why can't we choose the initial observer's conditions equal to the system's initial conditions?

Answer
We would love to, but in most cases we do not know the system's initial conditions.

Question
What constraint does not allow choosing the observer's poles far away to the left in the complex domain?

Answer
This would be a physical constraint on the system's bandwidth.

Solved Problems

Problem 4.1
Let the system be defined by

$$\dot{x} = \begin{pmatrix} 0 & 1 \\ 1 & 0 \end{pmatrix} x + \begin{pmatrix} 0 \\ 1 \end{pmatrix} u$$

$$y = (1 \quad 0)x$$

A. Check stability, controllability, and observability.
B. Design the controller with poles at $-1 \pm j$.
C. Design the observer with poles at $-2 \pm 2j$.

Solution
A. The characteristic polynomial in open loop is $\det(sI - A) = \det\begin{pmatrix} s & -1 \\ -1 & s \end{pmatrix} = s^2 - 1$; thus, the system's eigenvalues are $s_{1,2} = \pm 1$. Since we have a pole at the right complex semi-plane, the system is not stable. The controllability and observability are tested using the rank of controllability and observability matrices:

$$\mathcal{C} = \begin{pmatrix} 0 & 1 \\ 1 & 0 \end{pmatrix} \rightarrow \text{rank}(\mathcal{C}) = 2$$

$$\mathcal{O} = \begin{pmatrix} 1 & 0 \\ 0 & 1 \end{pmatrix} \rightarrow \text{rank}(\mathcal{O}) = 2$$

Therefore, the system is controllable and observable.

B. To design the controller with poles at $s_{1,2} = -1 \pm j$, we could compare the coefficients of $\alpha(s)$ and $a_k(s)$:

$$\begin{cases} \alpha(s) = (s-s_1)(s-s_2) = (s+1+j)(s+1-j) = s^2 + 2s + 2 \\ a_k(s) = \det(sI - A + BK) = \det\begin{pmatrix} s & -1 \\ -1+k_1 & s+k_2 \end{pmatrix} = s^2 + k_2 s + (k_1 - 1) \end{cases}$$

From the comparison of the coefficients, we get the following system of equations:

$$\begin{cases} 2 = k_2 \\ 2 = k_1 - 1 \end{cases}$$

which is solved by $k_1 = 3$ and $k_2 = 2$, or $K = (3 \quad 2)$.

C. To design the observer gains $L = \begin{pmatrix} l_1 \\ l_2 \end{pmatrix}$ with the poles at $\tilde{s}_{1,2} = -2 \pm 2j$, we compare the coefficients of $\tilde{\alpha}(s)$ and $a_L(s)$:

$$\begin{cases} \tilde{\alpha}(s) = (s-\tilde{s}_1)(s-\tilde{s}_2) = (s+2+2j)(s+2-2j) = s^2 + 4s + 8 \\ a_L(s) = \det(SI - A + LC) = \det\begin{pmatrix} s+l_1 & -1 \\ l_2-1 & s \end{pmatrix} = s^2 + l_1 s + l_2 - 1 \end{cases}$$

which produces the system of equations $4 = l_1$ and $8 = l_2 - 1$ solved by $L = \begin{pmatrix} 4 \\ 9 \end{pmatrix}$.

State Estimation (Observers)

Obviously, parts (B) and (C) could be solved by other methods using the Bass-Gura or the Ackermann formulas.

Now, let's simulate the system above in open loop (Figure 4.5a), closed loop with state-space controller where all states are measured by sensors (Figure 4.5b), and observer in closed loop with the same controller.

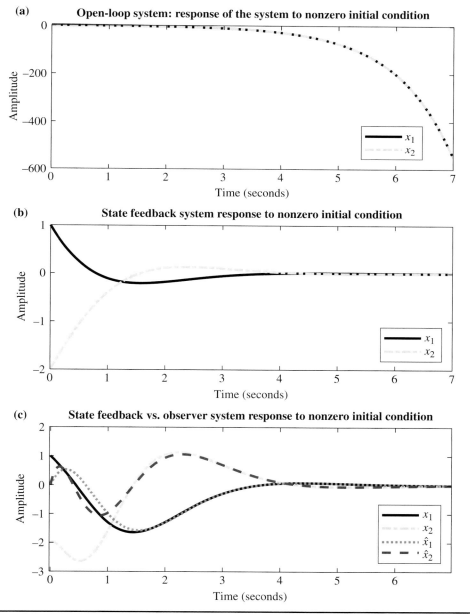

FIGURE 4.5 System simulation. (*a*) Open loop; (*b*) state feedback control; (*c*) state feedback control with observer.

As you can see, the open-loop system is diverging since it is unstable (Figure 4.5a), but the system in closed loop is converging to state zero for zero input from some initial conditions $x_0 = \begin{pmatrix} 1 \\ -2 \end{pmatrix}$. The convergence to small enough values takes about 4 seconds. Note that the convergence takes longer when the observer is present. It takes $\hat{x}$ about 2 seconds to approach from zero initial conditions to the real state x, and additional 3 seconds converging to small enough values of state. So, the overall convergence is slower, but it is promised not to diverge if the system is controllable and observable.

Problem 4.2
The discrete-time system is given by
$$x[k+1] = Ax[k] + Bu[k]$$
$$y[k] = Cx[k]$$

where $A = \begin{pmatrix} -1 & 1 & 0 \\ 1 & 0 & -1 \\ 0 & 1 & -1 \end{pmatrix}$; $B = \begin{pmatrix} 1 \\ 0 \\ 0 \end{pmatrix}$; $C = (0 \quad 0 \quad \eta)$, $0 < \eta < 1$.

A. Check controllability and observability of this system.
B. Design the observer's gain L with all the poles at η.
C. The controller's gain K is designed with all the poles at $(1-\eta)^2$. For which η values are the closed-loop poles slower than the observer's poles?

Solution
A. The controllability matrix $\mathcal{C} = \begin{pmatrix} 1 & -1 & 2 \\ 0 & 1 & -1 \\ 0 & 0 & 1 \end{pmatrix}$ is upper triangular with no zero values on the main diagonal; thus, it is full rank (rank($\mathcal{C}$) = 3), and the system is controllable.

The observability matrix $\mathcal{O} = \begin{pmatrix} 0 & 0 & \eta \\ 0 & \eta & -\eta \\ \eta & -\eta & 0 \end{pmatrix}$ is having three linearly independent rows for $\eta \neq 0$; thus, rank($\mathcal{O}$) = 3, and the system is observable.

B. Since the system is observable, it is possible to design an observer with poles anywhere we want, including at η. The desired characteristic polynomial would be $\tilde{\alpha}(z) = (z-\eta)^3 = z^3 - 3z^2\eta + 3z\eta^2 - \eta^3$. The inverse observability matrix is

$$\mathcal{O}^{-1} = \begin{pmatrix} 0 & 0 & \eta \\ 0 & \eta & -\eta \\ \eta & -\eta & 0 \end{pmatrix}^{-1} = \begin{pmatrix} 1/\eta & 1/\eta & 1/\eta \\ 1/\eta & 1/\eta & 0 \\ 1/\eta & 0 & 0 \end{pmatrix}$$

Using Ackermann's formula, we obtain the solution:

$$L = \tilde{\alpha}(A)\mathcal{O}^{-1}\begin{pmatrix} 0 \\ 0 \\ 1 \end{pmatrix}$$

$$= \left[\begin{pmatrix} -1 & 1 & 0 \\ 1 & 0 & -1 \\ 0 & 1 & -1 \end{pmatrix}^3 - 3\eta \begin{pmatrix} -1 & 1 & 0 \\ 1 & 0 & -1 \\ 0 & 1 & -1 \end{pmatrix}^2 + 3\eta^2 \begin{pmatrix} -1 & 1 & 0 \\ 1 & 0 & -1 \\ 0 & 1 & -1 \end{pmatrix} - \eta^3 \begin{pmatrix} 1 & 0 & 0 \\ 0 & 1 & 0 \\ 0 & 0 & 1 \end{pmatrix}\right]$$

$$\begin{pmatrix} 1/\eta & 1/\eta & 1/\eta \\ 1/\eta & 1/\eta & 0 \\ 1/\eta & 0 & 0 \end{pmatrix}\begin{pmatrix} 0 \\ 0 \\ 1 \end{pmatrix}$$

$$= \begin{pmatrix} -(\eta^3 + 3\eta^2 + 6\eta + 3)/\eta \\ (3\eta^2 + 3\eta + 1)/\eta \\ -(3\eta + 2)/\eta \end{pmatrix}$$

As you may have noticed, this task is very time consuming to do by hand. Alternative solutions with the Bass-Gura formula and the comparison of coefficients are long as well and preferably would be done with the help of computer.

C. For discrete-time systems the poles are "faster" if they are closer to the origin inside the unit circle. Thus, we require $|(1-\eta)^2| > |\eta|$, but since both sides are positive by the definition of η range $0 < \eta < 1$, we could remove the absolute value and solve the inequality $(1-\eta)^2 > \eta$ which is equivalent to $\eta^2 - 3\eta + 1 > 0$. This quadratic polynomial has two roots 0.382 and 2.618 and the parabola describing it is positive for $\eta < 0.382$ or for $\eta > 2.618$. Since $0 < \eta < 1$, the intersection of this interval with the inequality solution above is $0 < \eta < 0.382$.

Problem 4.3

The system is given in the following form:

$$\begin{cases} \ddot{x}(t) = Ax(t) + Bu(t) \\ y(t) = Cx(t) \end{cases}$$

Pay attention that the state variable x has the second-order time derivative. The matrix A is of the order n. Also, the pair $\{A, B\}$ is controllable, and the pair $\{A, C\}$ is observable.

A. What is the system's order in the standard state-space representation? Find the state vector $z(t)$ and matrices $\tilde{A}, \tilde{B}, \tilde{C}$ such that

$$\begin{cases} \dot{z}(t) = \tilde{A}z(t) + \tilde{B}u(t) \\ y(t) = \tilde{C}z(t) \end{cases}$$

B. Is the system controllable? Is it observable? Prove your answer for any order n.

For the rest of this question, assume $n = 1$, $B = \dfrac{1}{C}$, $A = a^2 \neq 0$.

C. Find the transfer function of the system. Is the system asymptotically stable?

D. Is it possible to stabilize the system with output feedback controller only (i.e., $u = -ky = -kcx$ with scalar k)? If it is possible, find such a controller k.

E. Is it possible to stabilize the system with the state feedback controller and observer? If yes, then design such a controller and observer.

Solution

A. We can define $z_1 = x$ and $z_2 = \dot{x}$ to express the new system; thus, the system's total order is $2n$, and

$$\begin{cases} \dot{z} = \begin{pmatrix} \dot{x} \\ \ddot{x} \end{pmatrix} = \underbrace{\begin{pmatrix} 0 & I_{n \times n} \\ A & 0 \end{pmatrix}}_{\tilde{A}} \underbrace{\begin{pmatrix} x \\ \dot{x} \end{pmatrix}}_{z} + \underbrace{\begin{pmatrix} 0 \\ B \end{pmatrix}}_{\tilde{B}} u \\ y = \underbrace{(C \quad 0)}_{\tilde{C}} \underbrace{\begin{pmatrix} x \\ \dot{x} \end{pmatrix}}_{z} \end{cases}$$

B. For $n \geq 2$, the controllability matrix is

$$\mathcal{C} = (\tilde{B} \quad \tilde{A}\tilde{B} \quad \tilde{A}^2\tilde{B} \quad \ldots \quad \tilde{A}^{2n-1}\tilde{B}) = \begin{pmatrix} 0 & B & 0 & AB & \ldots & 0 & A^{n-1}B \\ B & 0 & AB & 0 & \ldots & A^{n-1}B & 0 \end{pmatrix}$$

All even columns are linearly independent, since $\{A, B\}$ is a controllable pair. For the same reason, the odd columns are linearly independent. Moreover, the odd and even columns are independent; thus, matrix $\mathcal{C}$ has full $2n$ rank, and the system is controllable. Similarly, the rank of observability matrix is 2; thus, the system is observable.

C. Now,

$$\begin{cases} \dot{z} = \begin{pmatrix} 0 & 1 \\ a^2 & 0 \end{pmatrix} z + \begin{pmatrix} 0 \\ B \end{pmatrix} u \\ y = (B^{-1} \quad 0) z \end{cases}$$

Thus,

$$G(s) = \tilde{C}(sI - \tilde{A})^{-1}\tilde{B} = \frac{1}{s^2 - a^2}$$

The system has two poles at $\pm a$, so one of them has to be on the "right unstable area." Thus, the system is not stable.

D. Let's try to stabilize the system with a constant output controller $u = -kcx$. Then in case of a closed loop, we get

$$\dot{z} = (A - BkC)z = \begin{pmatrix} 0 & 1 \\ a^2 - k & 0 \end{pmatrix} z$$

The characteristic closed-loop polynomial is $s^2 - (a^2 - k)$; thus, the closed-loop system has poles at $\pm\sqrt{a^2 - k}$. One of those poles has to be unstable; thus, the system cannot be stabilized with such a controller.

E. We have proven in (B) that the system is controllable and observable; thus, it can be stabilized with the state feedback controller and observer.

Let's design the controller to get the closed-loop characteristic polynomial $\alpha(s) = (s+a)^2$.

$$a_k(s) = \det(sI - A + BK) = \begin{vmatrix} s & -1 \\ Bk_1 - a^2 & s + Bk_2 \end{vmatrix} = s^2 + Bk_2 s + Bk_1 - a^2$$

From the comparison of the coefficients: $K = (k_1 \quad k_2) = \dfrac{2a}{B}(a \quad 1)$.

Let's design an observer with all of the poles at $-4a$. We get $\alpha(s) = (s+4a)^2 = s^2 + 8as + 16a^2$, and

$$a_L(s) = \det(sI - A + LC) = \begin{vmatrix} s + l_1 B^{-1} & -1 \\ l_2 B^{-1} - a^2 & s \end{vmatrix} = s^2 + \dfrac{l_1}{B}s + \dfrac{l_2}{B} - a^2$$

After comparing the coefficients, $L = \begin{pmatrix} l_1 \\ l_2 \end{pmatrix} = Ba \begin{pmatrix} 8 \\ 17a \end{pmatrix}$.

CHAPTER 5
Nonminimal Canonical Forms

So far we have discussed how to design a state-space controller for controllable systems. When we needed to estimate states, we assumed that the system was observable. What if your system is not controllable or not observable? There is no need to panic, yet, because in many cases it is still possible to control your system successfully. In this chapter, first we will learn to deal with noncontrollable or nonobservable systems. We will then introduce the terms detectability and stabilizability. Finally, we will show how to design controller and observer for such systems.

Canonical Noncontrollable Form

Let $\{A, B, C\}$ be a system of the order n, with the controllability matrix $\mathcal{C}$, where rank($\mathcal{C}$) = $r < n$. Obviously, this system is not controllable since it has r out of n controllable eigenvalues. There exists the similarity transformation T such that

$$x = T\bar{x}$$

$$\bar{A} = T^{-1}AT = \begin{pmatrix} \bar{A}_e & \bar{A}_{12} \\ 0 & \bar{A}_{\text{not }e} \end{pmatrix} \begin{matrix} \}r \\ \}n-r \end{matrix}$$

$$\bar{B} = T^{-1}B = \begin{pmatrix} \bar{B}_e \\ 0 \end{pmatrix} \begin{matrix} \}r \\ \}n-r \end{matrix}$$

$$\bar{C} = CT = \begin{pmatrix} \bar{C}_1 & \bar{C}_2 \end{pmatrix}$$

(5.1)

This representation is called canonical noncontrollable form. Matrix $\bar{A}$ has four blocks. The top-left block size is $r \times r$ and together with the upper part of vector $\bar{B}$ it creates a controllable pair $\{\bar{A}_e, \bar{B}_e\}$. This also means that the eigenvalues of the matrix $\bar{A}_{\text{not }e}$ are not controllable.

How to Find This Transformation

Denote $T = \left(\underbrace{\boxed{T_1}}_{r} \ \underbrace{\boxed{T_2}}_{n-r} \right)$, then

$$T_1 = \underbrace{(B \ \ AB \ \ \ldots \ \ A^{r-1}B)}_{r \text{ columns of } \mathcal{C}} \mathcal{C}^{-1}\{\overline{A}_e, \overline{B}_e\} \tag{5.2}$$

where $\mathcal{C}^{-1}\{\overline{A}_e, \overline{B}_e\}_{r \times r}$ is the invertible matrix that can be chosen arbitrarily, and it will be the inverse controllability matrix of submatrices $\overline{A}_e$ and $\overline{B}_e$. It is convenient to choose this matrix as $I_{r \times r}$.

T_2 is a completion to a basis of T_1 (additional $n-r$ column vectors linearly independent from the T_1 columns).

Notes

1. The first r columns of the controllability matrix are linearly independent.
2. Denote $\overline{x} = \left(\begin{array}{c} \boxed{\overline{x}_1} \}r \\ \boxed{\overline{x}_2} \}n-r \end{array} \right)$, then $\{\overline{A}_e, \overline{B}_e\}$ is a controllable pair ($\overline{x}_1$ is controllable), and $\overline{x}_2$ is not controllable, since these state variables depend on themselves only and are not changed by the input [carefully check the structure of the matrices in (5.1) to understand why].
3. The transfer function has only controllable eigenvalues: $G(s) = C(sI - A)^{-1}B = \overline{C}_1(sI - \overline{A}_e)^{-1}\overline{B}_e$.
4. The system's eigenvalues consist of those of $\overline{A}_e$ (controllable) and $\overline{A}_{\text{not }e}$ (not controllable). In other words, the eigenvalues of $\overline{A}_e$ and $\overline{A}_{\text{not }e}$ together determine the eigenvalues of the original matrix A.

Canonical Nonobservable Form

Let $\{A, B, C\}$ be a system of the order n, with the observability matrix $\mathcal{O}$, where $\text{rank}(\mathcal{O}) = l < n$. There exists the similarity transformation T such that

$$x = T\overline{x}$$

$$\overline{A} = T^{-1}AT = \left(\begin{array}{cc} \underbrace{\boxed{\overline{A}_\mathcal{O}}}_{l} & \underbrace{\boxed{0}}_{n-l} \}l \\ \underbrace{\boxed{\overline{A}_{21}}}_{l} & \underbrace{\boxed{\overline{A}_{\text{not }\mathcal{O}}}}_{n-l} \}n-l \end{array} \right)$$

$$\overline{B} = T^{-1}B = \left(\begin{array}{c} \boxed{\overline{B}_1} \}l \\ \boxed{\overline{B}_2} \}n-l \end{array} \right) \tag{5.3}$$

$$\overline{C} = CT = \left(\underbrace{\boxed{\overline{C}_\mathcal{O}}}_{l} \ \underbrace{\boxed{0}}_{n-l} \right)$$

This representation is called canonical nonobservable form. The matrix $\bar{A}$ has four blocks. The top-left block size is $l \times l$, and together with the left part of vector $\bar{C}$ it creates an observable pair $\{\bar{A}_O, \bar{C}_O\}$. This also means that the eigenvalues of the matrix $\bar{A}_{not\,O}$ are not observable.

How to Find This Transformation

Denote $T^{-1} = \begin{pmatrix} \boxed{T_1} \}^l \\ \boxed{T_2} \}^{n-l} \end{pmatrix}$, then

$$T_1 = \mathcal{O}^{-1}\{\bar{A}_O, \bar{C}_O\} \begin{pmatrix} C \\ CA \\ \ldots \\ CA^{l-1} \end{pmatrix} \tag{5.4}$$

where $\mathcal{O}^{-1}\{\bar{A}_O, \bar{C}_O\}_{l \times l}$ is the invertible matrix that can be chosen arbitrarily, and it will be the inverse observability matrix of submatrices $\bar{A}_O$ and $\bar{C}_O$. It is convenient to choose this matrix as $I_{r \times r}$.

T_2 is a completion to a basis of T_1 (additional $n-l$ row vectors linearly independent from the T_1 rows).

NOTE The first l rows of the observability matrix are always linearly independent.

NOTE It is impossible to move eigenvalues that are not controllable with a state feedback controller, and it is impossible to control the state estimation rate of eigenvalues that are not observable.

Stabilizability and Detectability

A system is called *stabilizable* if all its noncontrollable eigenvalues are stable. A system is called *detectable* if all its nonobservable eigenvalues are stable.

Theorem 5.1

To stabilize a system with the state feedback and an observer in the closed loop, the system should be stabilizable and detectable.

How to Check Controllability and Observability of Eigenvalues

As we saw earlier, it is important to know which eigenvalues are not controllable or not observable. The most general method to check it is to find the transformation to a canonical noncontrollable or nonobservable form. Then, the eigenvalues of $\bar{A}_e$ are controllable, and those of $\bar{A}_{not\,e}$ are not controllable. The eigenvalues of $\bar{A}_O$ are observable, and those of $\bar{A}_{not\,O}$ are not observable.

There are other methods to check the same thing:

1. If matrix A is diagonal, each zero in vector B is accordant to a noncontrollable eigenvalue and each zero in C is accordant to a nonobservable eigenvalue.
2. If $\det(s_i I - A + BK) = 0$ for all K, then eigenvalue s_i is noncontrollable.

3. If $\det(s_i I - A + LC) = 0$ for all L, then eigenvalue s_i is nonobservable.
4. If one of the state equations looks like $\dot{x}_i = \alpha x_i (\alpha \in \mathbb{R})$ (some x_i variable depends on itself only, and not on the inputs), then the eigenvalue α is noncontrollable.
5. You may use combinatorial considerations: for example, if the system is of the order 3, and $\text{rank}(\mathcal{C}) = \text{rank}(\mathcal{O}) = 1$, then one of the eigenvalues is definitely noncontrollable and nonobservable.

Question
What can you say about the observability and controllability of the eigenvalues of a system with transfer function 0?

Answer
The system could be controllable, or observable, but not both. The simplest case where the transfer function equals zero is when vector B or C equals zero, but they don't have to be zero. You will find more information on this in solved Problem 5.2.

Question
What can you say about the stabilizability and detectability of the continuous-time system that has all its eigenvalues in the left half-plane?

Answer
This system is stabilizable and detectable, because all its noncontrollable and nonobservable eigenvalues are stable.

Question
How would you check the stabilizability of a multiple-input multiple-output (MIMO) system?

Answer
The stabilizability of a MIMO system is verified similar to single-input single-output (SISO) systems. If noncontrollable eigenvalues of a MIMO system are stable, then the system is stabilizable.

Solved Problems

Problem 5.1
The system is given by

$$\dot{x}(t) = Ax(t) + Bu(t)$$

$$y(t) = Cx(t)$$

$$A = \begin{pmatrix} 3 & 5 & 0 \\ 0 & -2 & 0 \\ -4 & -5 & -1 \end{pmatrix}; \quad B = \begin{pmatrix} 0 \\ 1 \\ 0 \end{pmatrix}; \quad C = \begin{pmatrix} 1 & 2 & 1 \end{pmatrix}$$

A. Check controllability and observability.
B. Check stabilizability and detectability.

C. Is it possible to stabilize the system with state feedback? If yes, design the controller.

D. Is it possible to stabilize the system with state feedback and observer?

Solution

A. First, we will compute the rank of the controllability and observability matrices to find out how many eigenvalues are not controllable or not observable:

$$\mathcal{C} = \begin{pmatrix} 0 & 5 & 5 \\ 1 & -2 & 4 \\ 0 & -5 & -5 \end{pmatrix}$$

Note that the first and the third rows are proportional (the third row is the first row times -1); thus, they are linearly dependent and the matrix rank is lower than 3. On the other hand, the second row cannot be written as a linear combination of other rows; thus, we have at least two linearly independent rows and the rank of $\mathcal{C}$ is at least 2. If the rank is at least 2 and less than 3, it must be 2. We conclude that two eigenvalues are controllable and one is not, but we don't know which one yet.

Similarly, the observability matrix is

$$\mathcal{O} = \begin{pmatrix} 1 & 2 & 1 \\ -1 & -4 & -1 \\ 1 & 8 & 1 \end{pmatrix}$$

The first and the last columns are equal, but the second column is not proportional to the first and the third columns; thus, there are two linearly independent columns and rank($\mathcal{C}$) = 2. Again, only one out of three eigenvalues is not observable, but we don't know which one.

B. To check stabilizability and detectability, we need to identify if the eigenvalue that is not controllable and the eigenvalue that is not observable are stable.

A standard way to solve this problem is to convert our system into a canonical noncontrollable or not observable form. To find the transformation to a canonical noncontrollable form, we take the first two columns of the controllability matrix [since rank($\mathcal{C}$) = 2] and add one more column in such a way that the transform matrix will be invertible. Obviously, there are infinite number of ways to do that, but we need to pick something simple like

$$T = \begin{pmatrix} 0 & 5 & 0 \\ 1 & -2 & 0 \\ 0 & -5 & 1 \end{pmatrix}$$

Thus, the transformed system is (after painfully long-hand computations or by using a computer)

$$\bar{A} = T^{-1}AT = \begin{pmatrix} 0 & 6 & 0 \\ 1 & 1 & 0 \\ 0 & 0 & -1 \end{pmatrix}$$

$$\bar{B} = T^{-1}B = \begin{pmatrix} 1 \\ 0 \\ 0 \end{pmatrix}$$

$$\bar{C} = CT = (2 \quad -4 \mid 1)$$

Note that the number of zeros in vector B is not defining the block size. In the transformed system, matrix $\bar{A}$ is block-diagonal; thus, its eigenvectors are the eigenvectors of the blocks on the main diagonal, namely $\begin{pmatrix} 0 & 6 \\ 1 & 1 \end{pmatrix}$ and (-1).

Since the bottom-right block in the canonical form is not controllable, the eigenvalue of the block (-1) is not controllable and cannot be moved with the state-space controller. The eigenvalue of this block has the value -1 itself and it is stable; thus, the system is stabilizable. In the next part of this problem, we will see a simpler way of identifying stabilizability and designing a controller at the same time.

Note that the controllability matrix of the upper-left (controllable) blocks $\begin{pmatrix} 0 & 6 \\ 1 & 1 \end{pmatrix}$ and $\begin{pmatrix} 1 \\ 0 \end{pmatrix}$ is the identity matrix $I_{2 \times 2}$ as determined by Formula (5.2).

Now, to find the eigenvalue that is not observable, we find the transformation to a canonical nonobservable state by taking the first two rows of the observability matrix and completing it with the linearly independent row $(0, 0, 1)$ to make T invertible:

$$T^{-1} = \begin{pmatrix} 1 & 2 & 1 \\ -1 & -4 & -1 \\ 0 & 0 & 1 \end{pmatrix}$$

The canonical form will be

$$\bar{A} = T^{-1}AT = \left(\begin{array}{cc|c} 0 & 1 & 0 \\ -2 & -3 & 0 \\ \hline -5.5 & -1.5 & 3 \end{array} \right)$$

$$\bar{B} = T^{-1}B = \begin{pmatrix} 2 \\ -4 \\ 0 \end{pmatrix}$$

$$\bar{C} = CT = (1 \quad 0 \mid 0)$$

In this transformed system, which is lower block-triangular, the bottom-right block of matrix A includes the not observable eigenvalue 3. Since this value is not stable (not in the left semi-plane), the system is not detectable.

C. The system is stabilizable; thus, we can design a state-space controller. We should be careful though, because we cannot move one of the eigenvalues, namely $s = -1$. Thus, we have to choose one of the closed-loop poles at -1. Two other closed-loop poles could be anywhere else in the left semi-plane. Let us design a controller to locate poles at $-1, -2$, and -3. Using the comparison of coefficients from Chapter 3, we compute the desired closed-loop characteristic polynomial:

$$\alpha(s) = (s+1)(s+2)(s+3) = (s+1)(s^2 + 5s + 6)$$

The closed-loop characteristic polynomial as a function of K vector elements will be

$$a_K(s) = \det(sI - A + BK)$$

$$= \det\left(\begin{pmatrix} s & 0 & 0 \\ 0 & s & 0 \\ 0 & 0 & s \end{pmatrix} - \begin{pmatrix} 3 & 5 & 0 \\ 0 & -2 & 0 \\ -4 & -5 & -1 \end{pmatrix} + \begin{pmatrix} 0 \\ 1 \\ 0 \end{pmatrix}(k_1 \quad k_2 \quad k_3)\right)$$

$$= \begin{vmatrix} s-3 & -5 & 0 \\ k_1 & s+2+k_2 & k_3 \\ 4 & 5 & s+1 \end{vmatrix}$$

If we develop this determinant by the first line

$$\begin{vmatrix} s-3 & -5 & 0 \\ k_1 & s+2+k_2 & k_3 \\ 4 & 5 & s+1 \end{vmatrix} = (s-3)\det\begin{pmatrix} s+2+k_2 & k_3 \\ 5 & s+1 \end{pmatrix} + 5\det\begin{pmatrix} k_1 & k_3 \\ 4 & s+1 \end{pmatrix}$$

$$= (s-3)[(s+2+k_2)(s+1) - 5k_3] + 5[(s+1)k_1 - 4k_3]$$

$$= (s-3)(s+2+k_2)(s+1) - 5k_3(s-3) + 5k_1(s+1) - 20k_3$$

$$= (s-3)(s+2+k_2)(s+1) + 5k_1(s+1) - 5k_3 - 5k_3 s$$

$$= (s-3)(s+2+k_2)(s+1) + 5k_1(s+1) - 5k_3(s+1)$$

$$= (s+1)[(s-3)(s+2+k_2) + 5k_1 - 5k_3]$$

$$= (s+1)[s^2 + (-1+k_2)s + 5k_1 - 5k_3 - 6 - 3k_2]$$

Notice that the factor $(s+1)$, which we cannot change automatically, became a multiplier of the quadratic polynomial. This will always happen to systems that are not controllable. The characteristic eigenvalues that are not controllable will be factored out in $\det(sI - A + BK)$, and this is an alternative method to identify eigenvalues that are not controllable.

Now, since the original desired polynomial is $\alpha(s) = (s+1)(s^2 + 5s + 6)$, we need to compare the coefficients of the quadratic polynomials only, that is, $s^2 + 5s + 6 = s^2 + (-1+k_2)s + 5k_1 - 5k_3 - 6 - 3k_2$. Thus, $k_2 - 1 = 5$ and $5k_1 - 5k_3 - 6 - 3k_2 = 6$. The gain $k_2 = 6$, and for two other gains we have one degree of freedom. In other words, we can freely choose k_3 to be, say, $k_3 = 0$, and then $k_1 = 6$. To summarize, $K = (6, 6, 0)$.

D. The system is not detectable; thus, it is not possible to stabilize the system with an observer and a controller.

Problem 5.2

Let $\{A, B, C\}$ be a SISO system, where $\{A, B\}$ is a noncontrollable pair. Prove that there exists $C \neq 0$ such that the system's transfer function is zero.

Solution

Without loss of generality, assume that the system is given in a canonical noncontrollable form. This assumption does not limit generality, since we are always able to find similarity transform from any system that is not controllable to a canonical form, and their transfer functions are the same. Let's start with block matrices given as follows in the canonical form:

$$\dot{x} = \begin{pmatrix} A_{11} & A_{12} \\ 0 & A_{22} \end{pmatrix} x + \begin{pmatrix} B_1 \\ 0 \end{pmatrix} u \qquad y = (C_1 \quad C_2) x$$

Now, the transfer function is

$$G(s) = C(sI - A)^{-1} B = (C_1, C_2) \begin{pmatrix} sI - A_{11} & -A_{12} \\ 0 & sI - A_{22} \end{pmatrix}^{-1} \begin{pmatrix} B_1 \\ 0 \end{pmatrix} =$$

$$= (C_1, C_2) \begin{pmatrix} (sI - A_{11})^{-1} & (sI - A_{11})^{-1} A_{12} (sI - A_{22})^{-1} \\ 0 & (sI - A_{22})^{-1} \end{pmatrix} \begin{pmatrix} B_1 \\ 0 \end{pmatrix} =$$

$$= (C_1, C_2) \begin{pmatrix} (sI - A_{11})^{-1} B_1 \\ 0 \end{pmatrix} = C_1 (sI - A_{11})^{-1} B_1$$

We can choose any C_2 and $C_1 = 0$. The transfer function will be zero, despite the fact that C is not zero.

Problem 5.3
The system is given by

$$\dot{x}(t) = Ax(t) + Bu(t)$$
$$y(t) = Cx(t)$$

$$A = \begin{pmatrix} 1 & 0 & 0 \\ 0 & -1 & 0 \\ -2 & 2 & 1 \end{pmatrix}; \quad B = \begin{pmatrix} 1 \\ 0 \\ 1 \end{pmatrix}; \quad C = (1 \quad 0 \quad 0)$$

A. Check controllability and observability.

B. Check stabilizability and detectability.

C. Is it possible to stabilize the system with state feedback? If yes, design the controller.

D. Is it possible to stabilize the system with state feedback and an observer?

Solution

$$A = \begin{pmatrix} 1 & 0 & 0 \\ 0 & -1 & 0 \\ -2 & 2 & 1 \end{pmatrix}; \quad B = \begin{pmatrix} 1 \\ 0 \\ 1 \end{pmatrix}; \quad C = (1 \quad 0 \quad 0)$$

A. To check the controllability and observability we need to compute the relevant matrices: $\mathcal{C} = (B, AB, A^2 B) = \begin{pmatrix} 1 & 1 & 1 \\ 0 & 0 & 0 \\ 1 & -1 & -3 \end{pmatrix} \Rightarrow r = \text{rank}(\mathcal{C}) = 2$ (row of zeros is always dependent and the other two vectors are not proportional, thus independent) and the system is not controllable.

$$\mathcal{O} = (C, CA, CA^2)^T = \begin{pmatrix} 1 & 0 & 0 \\ 1 & 0 & 0 \\ 1 & 0 & 0 \end{pmatrix} \Rightarrow l = \text{rank}(\mathcal{O}) = 1 \text{ (only one nonzero column)}$$

and the system is not observable.

B. There are a few ways to solve this problem. We will start with the standard way of computing the transformation to canonical noncontrollable realization. In most cases you will not need to do that.

We need the first $r = 2$ columns of the controllability matrix and one additional column which will make all three columns linearly independent. One obvious choice is a vector with a nonzero second element (since there is a row of zeros in the matrix). As usual, we choose the matrix $\mathcal{C}^{-1}\{\overline{A}_e, \overline{B}_e\} = I$. The transform is as follows:

$$T = \begin{pmatrix} 1 & 1 & 0 \\ 0 & 0 & 1 \\ 1 & -1 & 0 \end{pmatrix}$$

and the transformed system is

$$\overline{A} = T^{-1}AT = \begin{pmatrix} 0 & -1 & 1 \\ 1 & 2 & -1 \\ 0 & 0 & -1 \end{pmatrix} \quad \overline{B} = T^{-1}B = \begin{pmatrix} 1 \\ 0 \\ 0 \end{pmatrix}$$

The eigenvalues of the block triangular matrix are sitting in the matrices on the main diagonal. Since the lower-right block is scalar, −1 is the eigenvalue. As you may see, the second zero in the transformed B matrix does not mean that two eigenvalues are not controllable. Only the last eigenvalue (−1; on the bottom-right block) is not controllable. This eigenvalue is "stable"; thus, the system is stabilizable.

As a side note, notice that the $\overline{A}_e = \begin{pmatrix} 0 & -1 \\ 1 & 2 \end{pmatrix}$ and $\overline{B}_e = \begin{pmatrix} 1 \\ 0 \end{pmatrix}$, thus $\mathcal{C}\{\overline{A}_e, \overline{B}_e\} = I^{-1} = I$ as expected.

The same problem can be solved in a much simpler way by the observation that in the original system $\dot{x}_2 = -x_2$, which means that this state variable depends on itself, and is not affected by the input u. The appropriate eigenvalue is the coefficient of this state, namely "stable" −1. Moreover, there is only one eigenvalue that is not controllable since the rank of controllability matrix is 2 (out of 3). Thus, we can conclude that the system is stabilizable.

To answer the question of detectability, we need to compute the system's eigenvalues. Of course, we can compute $\det(sI − A) = 0$ roots, but it can be done by observation: Matrix A is block (lower) triangular (with the block of two zeros on the top right). So, the eigenvalues are 1 and $\begin{pmatrix} -1 & 0 \\ 2 & 1 \end{pmatrix}$. The last 2×2 matrix is triangular; thus, the eigenvalues are 1 and −1. In conclusion, the system's eigenvalues are 1, 1, −1. Only one of them is "stable," and two of them are not observable; thus, from combinatorial standpoint, the system should not be detectable. This concludes the argument, but if you want to know which two specific eigenvalues are not controllable, just observe that the system is given in the canonical nonobservable form, and thus the eigenvalues 1 and −1 are not observable.

C. The system is stabilizable, so we can design the controller. Let's use the coefficient comparison technique:

$$\det(sI - A + BK) = \det\left(\begin{pmatrix} s-1 & 0 & 0 \\ 0 & s+1 & 0 \\ 2 & -2 & s-1 \end{pmatrix} + \begin{pmatrix} k_1 & k_2 & k_3 \\ 0 & 0 & 0 \\ k_1 & k_2 & k_3 \end{pmatrix}\right) =$$

$$\det\begin{pmatrix} s-1+k_1 & k_2 & k_3 \\ 0 & s+1 & 0 \\ 2+k_1 & -2+k_2 & s-1+k_3 \end{pmatrix} = (s+1)[(s-1+k_1)(s-1+k_3) - k_3(2+k_1)]$$

Note that noncontrollable eigenvalues will always create factors $(s - s_i)$ which are independent of K values (that could be taken out of parentheses). Now we need to decide on the desired closed-loop polynomial. Note that any desired polynomial should include the term $(s + 1)$, since the eigenvalue -1 is not controllable. After deciding on the pole locations, say $-1, -3, -5$, we solve the problem as done earlier.

D. The system is not detectable; therefore, the design of the observer is impossible.

Problem 5.4

The system is given by

$$\begin{cases} \dot{x} = Ax + Bu \\ y = Cx \end{cases}$$

$$A = \begin{pmatrix} 2 & 1 & 5 & 2 \\ 3 & 0 & 3 & 7 \\ 0 & 0 & -3 & 5 \\ 0 & 0 & 0 & -4 \end{pmatrix} \quad B = \begin{pmatrix} 1 \\ -3 \\ 0 \\ 0 \end{pmatrix} \quad C = (1 \ 0 \ 7 \ 9)$$

A. Find the transfer function before and after the poles/zeros cancelation.

B. Find all the uncontrollable eigenvalues. Is the system stabilizable?

C. Find the similarity transform T that converts the system into a not controllable canonical form. Choose the following structure for T: $T = \begin{pmatrix} T_{11} & 0 \\ 0 & I \end{pmatrix}$, where T_{11} has the lowest possible order.

Solution

A. Similar to Problem 5.2, we divide matrix A into four blocks 2×2 (note that A is block triangular), and matrices B, C into halves and get the following system:

$$\dot{x} = \begin{pmatrix} A_{11} & A_{12} \\ A_{21} & A_{22} \end{pmatrix} x + \begin{pmatrix} B_1 \\ B_2 \end{pmatrix} u \quad y = (C_1 \ C_2) x$$

$$G(s) = (C_1 \vdots C_2) \begin{pmatrix} sI - A_{11} & -A_{12} \\ 0 & sI - A_{22} \end{pmatrix}^{-1} \begin{pmatrix} B_1 \\ \cdots \\ 0 \end{pmatrix} =$$

$$= (C_1 \vdots C_2) \begin{pmatrix} (sI - A_{11})^{-1} & * \\ 0 & (sI - A_{22})^{-1} \end{pmatrix} \begin{pmatrix} B_1 \\ \cdots \\ 0 \end{pmatrix} =$$

$$= C_1(sI - A_{11})^{-1} B_1 = (1 \; 0) \begin{pmatrix} s-2 & -1 \\ -3 & s \end{pmatrix}^{-1} \begin{pmatrix} 1 \\ -3 \end{pmatrix} =$$

$$(1 \; 0) \begin{pmatrix} \frac{s}{s^2 - 2s + 3} & \frac{1}{s^2 - 2s + 3} \\ \frac{3}{s^2 - 2s + 3} & \frac{s-2}{s^2 - 2s + 3} \end{pmatrix} \begin{pmatrix} 1 \\ -3 \end{pmatrix} = \frac{s-3}{s^2 - 2s + 3} = \frac{1}{s+1} \underset{\text{before cancelations}}{=}$$

$$= \frac{(s-3)(s+3)(s+4)}{(s-3)(s+3)(s+4)(s+1)} = \frac{1}{s+1}$$

B. Observe that $\{A_{11}, b_1\}$ is not controllable: $\mathcal{C}\{A_{11}, b_1\} = \begin{pmatrix} 1 & -1 \\ -3 & 3 \end{pmatrix}$; rank $= 1 < 2$.

Thus, the eigenvalue $s = 3$ is not controllable. The additional eigenvalues that are not controllable are those of A_{22}, that is, $s = -3$ and $s = -4$ (because x_3 and x_4 depend on themselves only and are not affected by the input u).

The system has an unstable, not controllable eigenvalue $s = 3$; thus, it is not stabilizable and it is impossible to design the state feedback controller.

C. It is easy to see that $T = \begin{pmatrix} 1 & 0 & 0 & 0 \\ -3 & 1 & 0 & 0 \\ 0 & 0 & 1 & 0 \\ 0 & 0 & 0 & 1 \end{pmatrix}$, where the first column is B, and the rest is just the simplest completion to the basis.

CHAPTER 6
Linearization

So far we have worked with linear models only. Fortunately, the real world is not that simple. The real problems are nonlinear, and quite often to the extent that they can't even be approximated by a linear counterpart. In this chapter, we deal with more "nicely behaved" nonlinear systems that can be locally approximated by a linear system.

Equilibrium Points

Most nonlinear systems described by nonlinear differential equations can be brought to an explicit state-space form:

$$\begin{cases} \dot{x} = f(x, u) \\ y = g(x, u) \end{cases} \quad x = (x_1, \ldots, x_n)^T \tag{6.1}$$

where f and g are some nonlinear functions.

The *equilibrium points* $\{x_e, u_e\}$ are defined by the solutions of the equation $f(x_e, u_e) = 0$ [or $x_e(k+1) = x_e(k)$ for discrete-time systems]. The meaning of this kind of n-dimensional point is that if the state x gets into that point ($x = x_e$), it will stay there if no additional inputs, noises, and disturbances are introduced in the system.

Suppose the input changes a little bit from the equilibrium point by $\delta u(t)$, then we can write this as $u(t) = u_e + \delta u(t)$. Similarly, we denote small deviations of x and y from the equilibrium point by

$$x(t) = x_e + \delta x(t)$$

$$y(t) = y_e + \delta y(t)$$

NOTE The variables δx and δy mean a small change in x and y, and not a multiplication of δ by x or y.

When we substitute these definitions in state Equation (6.1), we get

$$\begin{cases} \dfrac{d}{dt}(x_e + \delta x) = \dot{\delta x} = f(x_e + \delta x, u_e + \delta u) \\ y_e + \delta y = g(x_e + \delta x, u_e + \delta u) \end{cases} \tag{6.2}$$

The first equality is true because x_e is a constant, and the derivative of a constant is zero. Now, we develop $f(x_e + \delta x, u_e + \delta u)$ with the Taylor series to get the approximate linear model for small disturbances around the equilibrium point, or equilibrium trajectory. The Taylor series formula for multivariable function $h(\tilde{x})$ around a point a is given by (see Appendix E.7):

$$h(\tilde{x} + a) = h(a) + (\nabla h(a))^T \tilde{x} + \frac{1}{2} \tilde{x}^T \nabla^2 h(a) \tilde{x} + \text{(higher-order terms)} \qquad (6.3)$$

where ∇ denotes a gradient (vector of partial derivatives of h) and ∇^2 denotes the Hessian matrix of second-order derivatives. We will ignore all terms that are above order one and use only $h(\tilde{x} + a) \approx h(a) + (\nabla h(a))^T \tilde{x}$. In our case, the function itself is multidimensional, that is, $f = (f_1, f_2, \ldots, f_n)^T$, so we need to write the Taylor series for each $f_i; i = 1, \ldots, n$:

$$f_i(\tilde{x} + a) \approx f_i(a) + (\nabla f_i(a))^T \tilde{x}$$

If $\tilde{x} = (\delta x, \delta u)^T$ and $a = (x_e, u_e)^T$, we get

$$f_i(x_e + \delta x, u_e + \delta u) \approx f_i(x_e, u_e) + (\nabla f_i(x_e, u_e))^T \begin{pmatrix} \delta x \\ \delta u \end{pmatrix}$$

$$= f_i(x_e, u_e) + \left(\frac{\partial f_i}{\partial x_1} \quad \cdots \quad \frac{\partial f_i}{\partial x_n} \quad \frac{\partial f_i}{\partial u} \right)\bigg|_{x_e, u_e} \begin{pmatrix} \delta x_1 \\ \vdots \\ \delta x_n \\ \delta u \end{pmatrix}$$

$$= f_i(x_e, u_e) + \delta x_1 \frac{\partial f_i}{\partial x_1}\bigg|_{x_e, u_e} + \delta x_2 \frac{\partial f_i}{\partial x_2}\bigg|_{x_e, u_e} + \cdots + \delta x_n \frac{\partial f_i}{\partial x_n}\bigg|_{x_e, u_e} + \delta u \frac{\partial f_i}{\partial u}\bigg|_{x_e, u_e} \qquad (6.4)$$

By definition, $f_i(x_e, u_e) = 0$, thus $f_i(x_e + \delta x, u_e + \delta u) \approx \delta x_1 \frac{\partial f_i}{\partial x_1}\bigg|_{x_e, u_e} + \delta x_2 \frac{\partial f_i}{\partial x_2}\bigg|_{x_e, u_e} + \cdots + \delta x_n \frac{\partial f_i}{\partial x_n}\bigg|_{x_e, u_e} + \delta u \frac{\partial f_i}{\partial u}\bigg|_{x_e, u_e}$.

Similarly, for $y_e + \delta y = g(x_e + \delta x, u_e + \delta u)$, we have

$$y_e + \delta y \approx g(x_e, u_e) + \left(\frac{\partial g}{\partial x_1} \quad \cdots \quad \frac{\partial g}{\partial x_n} \quad \frac{\partial g}{\partial u} \right)\bigg|_{x_e, u_e} \begin{pmatrix} \delta x_1 \\ \vdots \\ \delta x_n \\ \delta u \end{pmatrix}$$

$$= g(x_e, u_e) + \delta x_1 \frac{\partial g}{\partial x_1}\bigg|_{x_e, u_e} + \delta x_2 \frac{\partial g}{\partial x_2}\bigg|_{x_e, u_e} + \cdots + \delta x_n \frac{\partial g}{\partial x_n}\bigg|_{x_e, u_e} + \delta u \frac{\partial g}{\partial u}\bigg|_{x_e, u_e} \qquad (6.5)$$

Note that y_e is canceled since, by definition, $y_e = g(x_e, u_e)$.

Linearization

To summarize (6.4) and (6.5) in the matrix form

$$\begin{cases} \dfrac{d}{dt}(\delta x(t)) \approx A\delta x(t) + B\delta u(t) \\ \delta y(t) \approx C\delta x(t) + D\delta u(t) \end{cases}$$

$$A = \left.\dfrac{\partial f}{\partial x}\right|_{\substack{x=x_e \\ u=u_e}} = \begin{pmatrix} \dfrac{\partial f_1}{\partial x_1} & \cdots & \dfrac{\partial f_1}{\partial x_n} \\ \vdots & \ddots & \vdots \\ \dfrac{\partial f_n}{\partial x_1} & \cdots & \dfrac{\partial f_n}{\partial x_n} \end{pmatrix} \qquad B = \left.\dfrac{\partial f}{\partial u}\right|_{\substack{x=x_e \\ u=u_e}} = \begin{pmatrix} \dfrac{\partial f_1}{\partial u} \\ \vdots \\ \dfrac{\partial f_n}{\partial u} \end{pmatrix} \qquad (6.6)$$

$$C = \left.\dfrac{\partial g}{\partial x}\right|_{\substack{x=x_e \\ u=u_e}} = \begin{pmatrix} \dfrac{\partial g}{\partial x_1} & \cdots & \dfrac{\partial g}{\partial x_n} \end{pmatrix} \qquad D = \left.\dfrac{\partial g}{\partial u}\right|_{\substack{x=x_e \\ u=u_e}}$$

Theorem 6.1 (Lyapunov)
1. If all the eigenvalues λ_i of A have $Re\{\lambda_i\} < 0$, then x_e is a stable equilibrium point.
2. If at least one eigenvalue of A has $Re\{\lambda_i\} > 0$, then x_e is an unstable equilibrium point.
3. If all the eigenvalues of A have $Re\{\lambda_i\} \leq 0$, it is impossible to determine the stability.

NOTE Matrix A is the Jacobian matrix of the system.

Solved Problems

Problem 6.1
The forced Van der Pol oscillator equation is given by $\ddot{y}(t) - \mu(1 - y^2(t))\dot{y}(t) + y(t) = u(t)$, where $\mu \geq 0$ is the constant damping coefficient. The forced input is denoted by u, the output by y, and the state variables are defined by $x_1 = y$; $x_2 = \dot{y}$.

  A. Write nonlinear state-space equations of the Van der Pol oscillator.
  B. Write linearized state-space equations of the Van der Pol oscillator for $u_e = 0$.
  C. Check the stability of the equilibrium point for $u_e = 0$.

Solution
A. We start with writing the first derivative of the state variables in terms of those variables and the input (using the differential equation given in the problem):

$$\dot{x}_1 = \dot{y} = x_2$$

$$\dot{x}_2 = \ddot{y} = \mu(1-y^2)\dot{y} - y + u = \mu(1-x_1^2)x_2 - x_1 + u$$

The output is given by y, thus by definition

$$y = x_1$$

We represented the system in the form (6.1) such that $g(x_1, x_2, u) = x_1$, and $f_1(x_1, x_2, u) = x_2$, and $f_2(x_1, x_2, u) = \mu(1 - x_1^2)x_2 - x_1 + u$.

B. First, we need to find the equilibrium point (x_e, u_e). To find this point, both derivatives of x_1 and x_2 should be zero. In other words, $f_1 = f_2 = 0$. Therefore, from the first-state equation, $x_2 = 0$, and if we substitute this into the second-state equation $\mu(1 - x_1^2)x_2 - x_1 + u_e = 0$, we get $x_1 = u_e$. Since $u_e = 0$, it follows that $x_1 = 0$.

Now, we use the formulas in (6.6) to linearize the system:

$$A = \left.\frac{\partial f}{\partial x}\right|_{\substack{x=x_e \\ u=u_e}} = \begin{pmatrix} \frac{\partial f_1}{\partial x_1} & \frac{\partial f_1}{\partial x_2} \\ \frac{\partial f_2}{\partial x_1} & \frac{\partial f_2}{\partial x_2} \end{pmatrix} = \begin{pmatrix} 0 & 1 \\ -2\mu x_1 x_2 - 1 & \mu(1 - x_1^2) \end{pmatrix} = \begin{pmatrix} 0 & 1 \\ -1 & \mu \end{pmatrix}$$

$$B = \left.\frac{\partial f}{\partial u}\right|_{\substack{x=x_e \\ u=u_e}} = \begin{pmatrix} \frac{\partial f_1}{\partial u} \\ \frac{\partial f_2}{\partial u} \end{pmatrix} = \begin{pmatrix} 0 \\ 1 \end{pmatrix}$$

$$C = \left.\frac{\partial g}{\partial x}\right|_{\substack{x=x_e \\ u=u_e}} = \begin{pmatrix} \frac{\partial g}{\partial x_1} & \frac{\partial g}{\partial x_2} \end{pmatrix} = \begin{pmatrix} 1 & 0 \end{pmatrix}$$

$$D = \left.\frac{\partial g}{\partial u}\right|_{\substack{x=x_e \\ u=u_e}} = 0$$

C. Based on the Lyapunov theorem, in order to check the stability we need to find the eigenvalues of matrix A. The characteristic polynomial is $\det(sI - A) = s^2 - tr(A)s + \det(A) = s^2 - \mu s + 1$. Since $\mu \geq 0$, the characteristic polynomial will always have roots in the right semi-plane (based on the Routh-Hurwitz criterion); thus, the system is unstable.

Problem 6.2
The system in Figure 6.1 consists of linear $\frac{2}{(s^2 + 15s)}$ and nonlinear $u(e) = e^2 - 16$ parts.

A. Write the differential equation connecting the reference r to the output y.
B. Define the state as $x_1 = y$ and $x_2 = \dot{y}$ and write the state equations.
C. Find the equilibrium points for a constant input $r(t) = 1$.
D. Find the linearization $\{A, B, C, D\}$ around all equilibrium points. Are the equilibrium points stable?

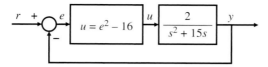

FIGURE 6.1 Nonlinear system example.

Solution

A. Note that we want to write the differential equation only in terms of r, y, and their derivatives. First, we can convert the linear part (transfer function) into a differential equation as explained in Chapter 1, Formula (1.11):

$$\frac{Y(s)}{U(s)} = \frac{2}{s^2 + 15s} \rightarrow (s^2 + 15s)Y(s) = s^2 Y(s) + 15sY(s) = 2U(s)$$

By applying the inverse Laplace transform to both sides, we get

$$\ddot{y}(t) + 15\dot{y}(t) = 2u(t)$$

Now, substituting u with the nonlinear equation $u = e^2 - 16$, we get $\ddot{y}(t) + 15\dot{y}(t) = 2(e^2 - 16)$. To find the error signal $e(t)$, we need to see how it is obtained in Figure 6.1. The negative feedback connection provides $e(t) = r(t) - y(t)$ which we can substitute back into our differential equation: $\ddot{y}(t) + 15\dot{y}(t) = 2(r(t) - y(t))^2 - 32$. So, the differential equation of interest will be

$$\ddot{y}(t) + 15\dot{y}(t) - 2(r(t) - y(t))^2 + 32 = 0$$

B. Given the state variable definitions $x_1 = y$ and $x_2 = \dot{y}$, we get the following state-space equations:

$$\begin{cases} \dot{x}_1 = \dot{y} = x_2 \\ \dot{x}_2 = -15x_2 + 2(r - x_1)^2 - 32 \\ y = x_1 \end{cases}$$

C. The equilibrium point should satisfy $\dot{x}_1 = \dot{x}_2 = 0$. If we solve the appropriate system of equations using the state equations we developed in (B) for $r(t) = 1$:

$$\dot{x}_1 = x_2 = 0 \rightarrow x_2 = 0$$

$$\dot{x}_2 = -15x_2 + 2(r - x_1)^2 - 32 = 0 \rightarrow (r - x_1)^2 = 16 \rightarrow r - x_1 = 4,$$

$$\text{or } r - x_1 = -4 \rightarrow x_1 = -3, \text{ or } x_1 = 5.$$

Thus, for $r = 1$, the equilibrium point is $x_{e_1} = (-3, 0)^T$ and $x_{e_2} = (5, 0)^T$.

D. In part (B) we have shown that $f_1 = x_2$; $f_2 = -15x_2 + 2(r - x_1)^2 - 32$; $g = x_1$; thus, for the first equilibrium point $x_{e_1} = (-3, 0)^T$,

$$A = \left.\frac{\partial f}{\partial x}\right|_{\substack{x=x_{e_1} \\ r=r_{e_1}}} = \begin{pmatrix} \frac{\partial f_1}{\partial x_1} & \frac{\partial f_1}{\partial x_2} \\ \frac{\partial f_2}{\partial x_1} & \frac{\partial f_2}{\partial x_2} \end{pmatrix} = \begin{pmatrix} 0 & 1 \\ -4(r-x_1) & -15 \end{pmatrix} = \begin{pmatrix} 0 & 1 \\ -16 & -15 \end{pmatrix}$$

$$B = \left.\frac{\partial f}{\partial r}\right|_{\substack{x=x_{e_1} \\ r=r_{e_1}}} = \begin{pmatrix} \frac{\partial f_1}{\partial r} \\ \frac{\partial f_2}{\partial r} \end{pmatrix} = \begin{pmatrix} 0 \\ 4(r-x_1) \end{pmatrix} = \begin{pmatrix} 0 \\ 16 \end{pmatrix}$$

$$C = \left.\frac{\partial g}{\partial x}\right|_{\substack{x=x_{e_1} \\ r=r_{e_1}}} = \begin{pmatrix} \frac{\partial g}{\partial x_1} & \frac{\partial g}{\partial x_2} \end{pmatrix} = (1 \quad 0)$$

$$D = \left.\frac{\partial g}{\partial r}\right|_{\substack{x=x_{e_1} \\ r=r_{e_1}}} = 0$$

The characteristic polynomial is $\det(sI - A) = s^2 + 15s + 16$. Since all coefficients are of the same sign (positive), all roots are stable. You can verify this directly by computing them ($s_1 = -1.16$ and $s_2 = -13.8$). Thus, based on the Lyapunov theorem, the first equilibrium point is stable.

For the second equilibrium point $x_{e_1} = (5, 0)^T$,

$$A = \left.\frac{\partial f}{\partial x}\right|_{\substack{x=x_{e_2} \\ r=r_{e_2}}} = \begin{pmatrix} \frac{\partial f_1}{\partial x_1} & \frac{\partial f_1}{\partial x_2} \\ \frac{\partial f_2}{\partial x_1} & \frac{\partial f_2}{\partial x_2} \end{pmatrix} = \begin{pmatrix} 0 & 1 \\ -4(r-x_1) & -15 \end{pmatrix} = \begin{pmatrix} 0 & 1 \\ 16 & -15 \end{pmatrix}$$

$$B = \left.\frac{\partial f}{\partial r}\right|_{\substack{x=x_{e_2} \\ r=r_{e_2}}} = \begin{pmatrix} \frac{\partial f_1}{\partial r} \\ \frac{\partial f_2}{\partial r} \end{pmatrix} = \begin{pmatrix} 0 \\ 4(r-x_1) \end{pmatrix} = \begin{pmatrix} 0 \\ -16 \end{pmatrix}$$

$$C = \left.\frac{\partial g}{\partial x}\right|_{\substack{x=x_{e_2} \\ r=r_{e_2}}} = \begin{pmatrix} \frac{\partial g}{\partial x_1} & \frac{\partial g}{\partial x_2} \end{pmatrix} = (1 \quad 0)$$

$$D = \left.\frac{\partial g}{\partial r}\right|_{\substack{x=x_{e_2} \\ r=r_{e_2}}} = 0$$

The characteristic polynomial is $\det(sI - A) = s^2 + 15s - 16$ and the roots $s_1 = 1$ and $s_2 = -16$. Based on the Lyapunov theorem, the second equilibrium point is not stable.

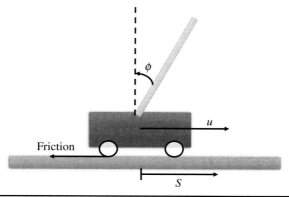

Figure 6.2 Inverted pendulum on a cart.

Problem 6.3
Inverted pendulum on a cart (see Figure 6.2) equations are given below:

$$(J + mL^2)\ddot{\phi} - mgL\sin\phi + mL\ddot{S}\cos\phi = 0$$
$$M\ddot{S} + F\dot{S} = u$$

where ϕ = rod angle
  g = gravitation acceleration constant
  S = linear translation
  u = external force applied to the cart
  M = cart mass
  F = friction coefficient
  J = moment of inertia around the rod's center of mass
  m = rod mass
  L = distance between the rod's center of mass and the axis

It is clear that the system is nonlinear, and since we have second-order derivatives of ϕ and S, this is a fourth-order system.

The numerical system parameters are given:

$$F = 1 \text{ kg}\cdot\text{s}^{-1}; \quad M = 1 \text{ kg}$$
$$g = 10 \text{ m}\cdot\text{s}^{-2}; \quad L = 0.5 \text{ m}$$
$$m = 0.5 \text{ kg}; \quad J = 0.125 \text{ kg}\cdot\text{m}^2$$

A. Linearize the system using the following state definition:

$$\begin{cases} x_1 = S \\ x_2 = \dot{S} \\ x_3 = S + \dfrac{J + mL^2}{mL}\phi \\ x_4 = \dot{S} + \dfrac{J + mL^2}{mL}\dot{\phi} \end{cases}$$

B. Is the system stable at its equilibrium point?

Chapter Six

Solution

A. Given that there are so many variables and constants, it is very important to understand what should stay in the state-space equations and what should disappear. Most of the letters we used to describe a model are constants; thus, they can stay. The variables, which are the functions of time, are S and ϕ. We want to replace these variables with the defined state variables $x_1, \ldots, x_4$. Given the chosen state variable x_3, $\phi = \dfrac{mL}{J+mL^2}(x_3 - S) = \dfrac{mL}{J+mL^2}(x_3 - x_1)$. The development of state equations is as follows:

$$\begin{cases} \dot{x}_1 = \dot{S} = x_2 \\ \dot{x}_2 = \ddot{S} \underset{\substack{\uparrow \\ \text{from the second} \\ \text{differential equation}}}{=} -\dfrac{F}{M}\dot{S} + \dfrac{1}{M}u = -\dfrac{F}{M}x_2 + \dfrac{1}{M}u \\ \dot{x}_3 = x_4 \\ \dot{x}_4 = \ddot{S} + \dfrac{J+mL^2}{mL}\ddot{\phi} = -\dfrac{F}{M}x_2 + \dfrac{1}{M}u + \dfrac{J+mL^2}{mL}\ddot{\phi} \end{cases}$$

Finally, we can use the first system's differential equation $(J+mL^2)\ddot{\phi} - mgL\sin\phi + mL\ddot{S}\cos\phi = 0$ to compute:

$$\ddot{\phi} = \dfrac{mgL}{J+mL^2}\sin\phi - \dfrac{mL}{J+mL^2}\ddot{S}\cos\phi$$

$$= \dfrac{mgL}{J+mL^2}\sin\left(\dfrac{mL}{J+mL^2}(x_3-x_1)\right) - \dfrac{mL}{J+mL^2}\ddot{S}\cos\left(\dfrac{mL}{J+mL^2}(x_3-x_1)\right) =$$

$$= \dfrac{mgL}{J+mL^2}\sin\left(\dfrac{mL}{J+mL^2}(x_3-x_1)\right) - \dfrac{mL}{J+mL^2}\left(-\dfrac{F}{M}x_2 + \dfrac{1}{M}u\right)\cos\left(\dfrac{mL}{J+mL^2}(x_3-x_1)\right)$$

If we substitute this expression in the fourth-state equation,

$$\dot{x}_4 = -\dfrac{F}{M}x_2 + \dfrac{1}{M}u + \dfrac{J+mL^2}{mL}\ddot{\phi} = -\dfrac{F}{M}x_2 + \dfrac{1}{M}u + \dfrac{J+mL^2}{mL}\left(\dfrac{mgL}{J+mL^2}\sin\left(\dfrac{mL}{J+mL^2}(x_3-x_1)\right)\right.$$

$$\left. -\dfrac{mL}{J+mL^2}\left(-\dfrac{F}{M}x_2 + \dfrac{1}{M}u\right)\cos\left(\dfrac{mL}{J+mL^2}(x_3-x_1)\right)\right)$$

$$= \dfrac{u-Fx_2}{M}\left(1-\cos\left(\dfrac{mL}{J+mL^2}(x_3-x_1)\right)\right) + g\sin\left(\dfrac{mL}{J+mL^2}(x_3-x_1)\right)$$

This way we have computed four state equations.

If we substitute the numerical values for the system's parameters, we get the following equations:

$$\begin{cases} \dot{x}_1 = x_2 \\ \dot{x}_2 = -x_2 + u \\ \dot{x}_3 = x_4 \\ \dot{x}_4 = (u - x_2)(1 - \cos(x_3 - x_1)) + 10\sin(x_3 - x_1) \end{cases}$$

Note that three of them are already linear; thus, linearization will not change them. The last equation is linearized around the equilibrium point $x_e = 0$:

$$\left.\frac{\partial \dot{x}_4}{\partial x_1}\right|_{x=0} = -10\cos(x_3 - x_1)\Big|_{x=0} = -10$$

$$\left.\frac{\partial \dot{x}_4}{\partial x_2}\right|_{x=0} = -(1 - \cos 0) = 0$$

$$\left.\frac{\partial \dot{x}_4}{\partial x_3}\right|_{x=0} = 10\cos(x_3 - x_1)\Big|_{x=0} = 10$$

$$\left.\frac{\partial \dot{x}_4}{\partial x_4}\right|_{x=0} = 0$$

Thus, $\dot{x}_4 = -10x_1 + 10x_3$.

Similarly,

$$\text{vector } B = \left(\frac{\partial \dot{x}_1}{\partial u} \quad \frac{\partial \dot{x}_2}{\partial u} \quad \frac{\partial \dot{x}_3}{\partial u} \quad \frac{\partial \dot{x}_4}{\partial u}\right)^T = (0 \quad 1 \quad 0 \quad (1 - \cos(x_3 - x_1))|_{x=0})^T = (0 \quad 1 \quad 0 \quad 0)^T.$$

B. Matrix $A = \begin{pmatrix} 0 & 1 & 0 & 0 \\ 0 & -1 & 0 & 0 \\ 0 & 0 & 0 & 1 \\ -10 & 0 & 10 & 0 \end{pmatrix}$, $\det(sI - A) = \cdots = s(s+1)(s+\sqrt{10})(s-\sqrt{10})$ and the system is unstable. Note that since matrix A is block-triangular, the eigenvalues of this matrix and the poles of the system are the same as eigenvalues of the main diagonal blocks $\begin{pmatrix} 0 & 1 \\ 0 & -1 \end{pmatrix}$ and $\begin{pmatrix} 0 & 1 \\ 10 & 0 \end{pmatrix}$.

CHAPTER 7
Lyapunov Stability

You are already familiar with asymptotic and bounded-input bounded-output (BIBO) stability. In this chapter, we will learn about additional type of stability which is appropriate not only for linear but also for nonlinear systems represented in state space. This stability was developed by Russian mathematician Alexandr Lyapunov in 1892 (Lyapunov, 1892). Lyapunov stability is weaker than asymptotic stability (which means that asymptotically stable systems are always Lyapunov stable), but many times the Lyapunov theory is the only way to analyze the stability of nonlinear systems, and it is still dominant in analyzing nonlinear systems. Lyapunov theorems also have theoretical importance in developing optimal control theory. It is advisable to read through Appendix G.20 before starting this chapter.

Internally Stable Systems

As we have seen in Chapter 6, the equilibrium points x_e are defined by the solutions of the equation $f(x_e) = 0$ (or $x_e(k+1) = x_e(k)$ for discrete-time systems).

Definition (Internal Stability in Continuous Time)

Let $\dot{x} = f(x, t)$, $x(t_0) = x_0$ be a state-space (possibly nonlinear) system with equilibrium point x_e. This equilibrium point is *internally stable* (Lyapunov stable) if for every $R > 0$, there exists $r(R) > 0$ such that if $\|x(0) - x_e\| < r$, then $\|x(t) - x_e\| \leq R$; $\forall t \geq t_0$, where $x(t)$ is the solution of the system at time t.

We can think about a specific state x as a point in an n-dimensional space. The change in state as a function of time is some trajectory in an n-dimensional space. The equilibrium point is Lyapunov stable if for any chosen distance R (from x_e) one can find a small enough distance r, such that all trajectories of the state starting within that small circle will not go farther than distance R (see Figure 7.1). Even simpler, state x that starts "close enough" to the equilibrium will stay "close enough" forever.

> **NOTE** The norm $\|\blacksquare\|$ could be thought of as the size of the vector or distance from the origin $x = 0$; thus, the norm of a difference of two vectors is the distance between those vectors in an n-dimensional space.

Definition (Asymptotic Stability)

The system is *asymptotically stable* if it is Lyapunov stable and if there exists $r > 0$ such that if $\|x(0) - x_e\| < r$, then $\lim_{t \to \infty} \|x(t) - x_e\| = 0$.

In other words, the system converges to its equilibrium point from "close enough" points.

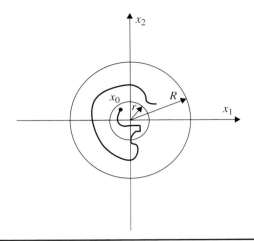

FIGURE 7.1 Lyapunov stability in 2D state space.

Direct Lyapunov Method (Second Method)

The definition of internal stability is very difficult to apply in practice because we need to compute the entire trajectory to make sure it is not leaving the region of interest. Now we will establish the stability property without explicit computation of the trajectories. The idea is that some sort of the system's energy dissipation rate teaches us something about the stability.

For the equilibrium point in the origin ($x_e = 0$), we will define a system's energy function. If such a function exists, then the system is stable.

Definition (Lyapunov Energy Function)

A function $V(x)$ (scalar function of the elements of the state vector x) is called *Lyapunov function* if

1. $V(x)$ is a continuous function with continuous partial derivatives with respect to all state variables (i.e., $\dfrac{dV}{dx_i}$ exists and is continuous).
2. $\forall x \neq 0 : V(x) > 0$.
3. $V(0) = 0$.
4. $\dot{V}(x) \leq 0$ for all x in the neighborhood of the origin.

To compute the derivative of V, we use the following:

$$\dot{V}(x) = \frac{\partial V(x)}{\partial t} = \begin{pmatrix} \dfrac{\partial V}{\partial x_1} & \dfrac{\partial V}{\partial x_2} & \cdots & \dfrac{\partial V}{\partial x_n} \end{pmatrix} f(x,t) = \frac{\partial V}{\partial x_1} f_1 + \cdots + \frac{\partial V}{\partial x_n} f_n \qquad (7.1)$$

where $\dot{x} = f(x,t)$.

NOTE Different linear or nonlinear systems will have different Lyapunov energy functions. If you have a function that will satisfy all four conditions, then you are lucky since the system is stable. If the function you have is not satisfying at least one of the conditions, it means nothing; the system could be stable or unstable and you need to search for another function. The first simple thing to try would be $V(x) = \alpha x_1^2 + \beta x_2^2 + \cdots$, but it is not promised to work. Unfortunately, standard formula for the Lyapunov energy function exists only for linear systems. It might be tremendously difficult to design Lyapunov energy functions for nonlinear systems and this topic is beyond the scope of this book.

Stability Theorem 7.1

1. The equilibrium point $x_e = 0$ is *Lyapunov stable* in its neighborhood if there exists Lyapunov function $V(x)$.
2. The equilibrium point $x_e = 0$ is *asymptotically stable* if it is Lyapunov stable and for all states x in the neighborhood (excluding x_e) exists: $\dot{V}(x) < 0$.

NOTES

1. The neighborhood means an n-dimensional hypersphere around the equilibrium point.
2. In stability Theorem 7.1, the only difference between Lyapunov and asymptotic stability is in the last (fourth) condition on the derivative of the Lyapunov energy function. The energy change (derivative) should be nonpositive for Lyapunov stability and strictly negative for asymptotic stability.
3. Even if we are unable to find energy function with the appropriate properties, this does not mean that the system is stable or unstable.
4. If the equilibrium point is not at the origin, the state equations can always be transformed in a way that this point will be moved to the origin. To do that, define new state variables $z = x - x_e$, and then $\dot{z} = f(z + x_e)$, and obviously $z_e = 0$. It is possible to explore the stability of the new equilibrium point z_e, and to conclude on the stability of the original x_e.

Lyapunov Stability for Continuous-Time LTI Systems

The linear time invariant (LTI) system $\dot{x} = Ax$ is *globally asymptotically stable* if from any initial condition x_0, $\lim_{t \to \infty} x(t) = 0$ (converging to zero state).

Lyapunov Theorem 7.2

The LTI system $\dot{x} = Ax$ is globally asymptotically stable if and only if for all A's eigenvalues s_i, $Re\{s_i\} < 0$.

Lyapunov Theorem 7.3

The LTI system $\dot{x} = Ax$ is globally asymptotically stable if for some positive definite symmetric matrix $P > 0$, the matrix $A^T P + PA$ is symmetric and negative definite.

Lyapunov Theorem 7.4
If the LTI system $\dot{x} = Ax$ is globally asymptotically stable and $Q > 0$ is a positive definite symmetric matrix, then exists $P > 0$ positive definite and symmetric such that the following *Lyapunov equation* takes place:

$$A^T P + PA = -Q \qquad (7.2)$$

Lyapunov Stability for Discrete-Time LTI Systems
Let $x[k+1] = Ax[k]$ be a discrete LTI system. This system is globally asymptotically stable if for all initial conditions x_0, $\lim_{k \to \infty} x[k] = 0$ (converging to zero state).

Lyapunov Theorem 7.5
The discrete-time LTI system $x[k+1] = Ax[k]$ is globally asymptotically stable if and only if for all A's eigenvalues s_i, $|s_i| < 1$.

Lyapunov Theorem 7.6
The discrete-time LTI system is globally asymptotically stable if for some positive definite symmetric matrix $P > 0$, the matrix $A^T PA - P$ is symmetric and negative definite.

Lyapunov Theorem 7.7
If the discrete-time LTI system is globally asymptotically stable and $Q > 0$ is a positive definite symmetric matrix, then exists $P > 0$ positive definite and symmetric such that the following *Lyapunov equation* takes place:

$$A^T PA - P = -Q \qquad (7.3)$$

NOTES

1. To check the stability of the LTI system, it is enough to pick one positive definite and symmetric matrix Q (say, $Q = I$), and to find the solution P of the Lyapunov equation. If the matrix P is positive definite and symmetric, then the system is globally asymptotically stable.
2. The appropriate Lyapunov function of the LTI system is $V(x) = x^T Px$.

Question
Can you think of a system which is Lyapunov stable but not asymptotically stable?

Answer
There are many such systems. For example, a car standing on a flat surface. If we move this car a little from its initial position (disturbance), it will not return to its initial position, but it will also not try to escape.

Question
Can you think of a system which is not Lyapunov stable, but from any initial condition (and zero input), $\lim_{t \to \infty} y(t) = 0$. Note that this kind of system is not asymptotically stable, though it always converges to zero state.

Answer
This system is more difficult to come by. Imagine a toy gun where the bullet is attached to the gun with a spring or a rubber band. From any shooting angle, the gun shoots the

bullet to a given distance and then the bullet returns to its original position using the spring. While the final position of the bullet is at the origin, it must leave the equilibrium neighborhood first.

Question
Is there any linear system which is asymptotically stable, but not globally asymptotically stable?

Answer
No, such a system does not exist. If the LTI system is stable for some initial conditions, then it is stable for any initial conditions. It is clear from Theorems 7.2 and 7.5 that asymptotic stability is dependent only on A's eigenvalues and not on initial conditions.

Question
Is there any linear system which has equilibrium points that are not in the origin?

Answer
Yes, for example $\begin{cases} \dot{x}_1 = x_1 \\ \dot{x}_2 = x_1 \end{cases}$ is obviously linear, but if $\dot{x}_1 = \dot{x}_2 = 0$, then $x_1 = 0$ and x_2 could be anything. So, we have an infinite number of equilibrium points. In general, for any linear system $\dot{x} = Ax$, if A is invertible, then the only equilibrium point is $x_e = 0$, but if A is singular, then there are an infinite number of solutions to $Ax = 0$, and hence an infinite number of equilibrium points.

Solved Problems

Problem 7.1
Given the system $\dot{x} = \begin{pmatrix} 0 & 1 \\ -1 & -2 \end{pmatrix} x + \begin{pmatrix} 0 \\ 1 \end{pmatrix} u$, analyze the system's stability using the Lyapunov equation.

Solution
We choose a positive definite identity matrix $Q = I$ and use Equation (7.2): $A^T P + PA = -I$.

$$\begin{pmatrix} 0 & -1 \\ 1 & -2 \end{pmatrix} \begin{pmatrix} p_1 & p_2 \\ p_2 & p_3 \end{pmatrix} + \begin{pmatrix} p_1 & p_2 \\ p_2 & p_3 \end{pmatrix} \begin{pmatrix} 0 & 1 \\ -1 & -2 \end{pmatrix} = \begin{pmatrix} -1 & 0 \\ 0 & -1 \end{pmatrix}$$

Note that the matrix P is symmetrical; thus, it needs only three free parameters: $p_1, p_2,$ and p_3. The system above also has only three independent linear equations (instead of four) because of symmetricity. If we multiply and add all matrices above, we get

$$\begin{cases} 0 \cdot p_1 - 1 \cdot p_2 + p_1 \cdot 0 + p_2 \cdot (-1) = -1 \\ 0 \cdot p_2 - 1 \cdot p_3 + p_1 \cdot 1 + p_2 \cdot (-2) = 0 \\ 1 \cdot p_1 - 2 \cdot p_2 + p_2 \cdot 0 + p_3 \cdot (-1) = 0 \\ 1 \cdot p_2 - 2 \cdot p_3 + p_2 \cdot 1 + p_3 \cdot (-2) = -1 \end{cases}$$

$$\begin{cases} -2p_2 = -1 \\ p_1 - 2p_2 - p_3 = 0 \\ p_1 - 2p_2 - p_3 = 0 \\ 2p_2 - 4p_3 = -1 \end{cases}$$

The solution of that linear system of equations is $p_1 = 1.5$; $p_2 = 0.5$; $p_3 = 0.5$. The resulting matrix is $P = \begin{pmatrix} 1.5 & 0.5 \\ 0.5 & 0.5 \end{pmatrix}$. We can use the Silvester criterion to prove that this matrix is positive definite. Both minors $|1.5| > 0$ and $\begin{vmatrix} 1.5 & 0.5 \\ 0.5 & 0.5 \end{vmatrix} = 0.25 > 0$ are positive; thus, the matrix is positive definite, and the system is stable. Alternatively, you could check eigenvalues and make sure that both eigenvalues of P are positive.

Problem 7.2
Given the system $\begin{cases} \dot{x}_1 = x_2 + u \\ \dot{x}_2 = -2x_1 - 2x_2 - 4x_1^3 \\ y = x_1 + x_2 \end{cases}$, prove the asymptotic stability of the origin.

Solution
The linearized system will be

$$\begin{cases} \dot{x}_1 = x_2 + u \\ \dot{x}_2 = -2x_1 - 2x_2 \\ y = x_1 + x_2 \end{cases}$$

Note that in this case, for the linearization around the origin we just need to ignore the higher-order terms (such as power of 3) and leave only the linear terms, because higher-order terms around zero will be very small compared to the first-order terms.

The system's matrix $A = \begin{pmatrix} 0 & 1 \\ -2 & -2 \end{pmatrix}$ and its characteristic polynomial is $\det(sI - A) = s^2 + 2s + 2$. Since all coefficients of that second-degree polynomial are positive, the system is asymptotically stable.

Problem 7.3
Given the discrete-time LTI system, prove that the eigenvalues of this system are inside a circle with the radius σ^{-1} if and only if for symmetric $Q > 0$ exists symmetric $P > 0$ such that $\sigma^2 A^T P A - P = -Q$.

Solution
Note that for $\sigma = 1$ we get Lyapunov's Theorem 7.7. We need to devise a plan for the solution. The idea is that we want to transform matrix A as a function of σ in a way that we will be able to use some Lyapunov theorem. Here is the plan:

1. We will prove that if matrix A has eigenvalues s_i, then matrix σA has eigenvalues σs_i.
2. From the stability of matrix σA we derive $|\sigma s_i| < 1$ (all eigenvalues in the unit circle).
3. The expression $|\sigma s_i| < 1$ is equivalent to $|s_i| < \frac{1}{\sigma} = \sigma^{-1}$ (eigenvalues s_i are in a circle of radius σ^{-1}) for positive σ.
4. Let's define $\tilde{A} = \sigma A$ and $\tilde{s}_i$ will be eigenvalues of $\tilde{A}$. The characteristic polynomial of $\tilde{A}$ is $\det(\tilde{s}_i I - \tilde{A}) = 0 \Leftrightarrow \det(\tilde{s}_i I - \sigma A) = 0 \Leftrightarrow \det(\sigma(\sigma^{-1}\tilde{s}_i I - A)) = 0 \Leftrightarrow \det(\sigma^{-1}\tilde{s}_i I - A) = 0$. This means that if $\tilde{s}_i$ is the eigenvalue of $\tilde{A}$ then $\sigma^{-1}\tilde{s}_i$ is the eigenvalue of A. This is equivalent to what we wanted to prove in part 1.

5. From Lyapunov's Theorem 7.7, for stable system there exist positive definite and symmetrical P and Q such that $\tilde{A}^T P \tilde{A} - P = -Q$. If we substitute the definition of $\tilde{A}$, $(\sigma A)^T P (\sigma A) - P = -Q$. For LTI systems this is true if and only if σA is stable. σA is stable if and only if $|\sigma s_i| < 1$, or alternatively $|s_i| < \sigma^{-1}$. This concludes the proof.

Problem 7.4
A system is given by the following nonlinear equations:

$$\begin{cases} \dot{x}_1 = -x_1 + 0.5 x_2 \sin(x_2) \\ \dot{x}_2 = -x_2 + 0.5 x_1 \cos(x_1) \end{cases}$$

Verify the Lyapunov stability of the origin equilibrium point using energy function $V = 0.5(x_1^2 + x_2^2)$.

Solution
Using Lyapunov's Theorem 7.1, let's compute the derivative of the given energy function using Formula (7.1): $\dot{V} = \dfrac{\partial V}{\partial x_1} f_1 + \cdots + \dfrac{\partial V}{\partial x_n} f_n$, where $\dot{x}_1 = f_1$ and $\dot{x}_2 = f_2$.

$$\dot{V} = x_1 \dot{x}_1 + x_2 \dot{x}_2 = x_1(-x_1 + 0.5 x_2 \sin(x_2)) + x_2(-x_2 + 0.5 x_1 \cos(x_1))$$
$$= -x_1^2 + 0.5 x_1 x_2 \sin(x_2) - x_2^2 + 0.5 x_1 x_2 \cos(x_1)$$
$$= -x_1^2 - x_2^2 + 0.5 x_1 x_2 (\sin(x_2) + \cos(x_1))$$

Now, knowledge of inequalities should help us find the solution. We want to prove that the expression above is never positive. We know that $0.5 x_1 x_2 (\sin(x_2) + \cos(x_1)) \leq |0.5 x_1 x_2 (\sin(x_2) + \cos(x_1))| = 0.5 |x_1 x_2| \cdot |\sin(x_2) + \cos(x_1)| \leq 0.5 |x_1 x_2|(|\sin(x_2)| + |\cos(x_1)|) \leq 0.5 |x_1 x_2|(1 + 1) = |x_1 x_2|$.

So we have, $\dot{V} \leq -x_1^2 - x_2^2 + |x_1 x_2|$, but $|x_1 x_2| \leq 0.5(x_1^2 + x_2^2)$, which is easily derived from the obvious inequality $(|x_1| - |x_2|)^2 \geq 0$. In conclusion, $\dot{V} \leq -0.5(x_1^2 + x_2^2) < 0$ for any $x_1 \neq 0$ and $x_2 \neq 0$. Therefore, the equilibrium point is stable.

Problem 7.5
Prove that the controllability Gramian $P = \int_0^\infty e^{A^T t} Q e^{At} \, dt$ solves the Lyapunov equation $A^T P + PA = -Q$ for stable A.

Solution

$$A^T P + PA = A^T \int_0^\infty e^{A^T t} Q e^{At} \, dt + \int_0^\infty e^{A^T t} Q e^{At} \, dt \, A = \int_0^\infty A^T e^{A^T t} Q e^{At} \, dt + \int_0^\infty e^{A^T t} Q e^{At} A \, dt$$

$$= \int_0^\infty A^T e^{A^T t} Q e^{At} + e^{A^T t} Q e^{At} A \, dt = \int_0^\infty \frac{d}{dt}(e^{A^T t} Q e^{At}) \, dt = e^{A^T t} Q e^{At} \Big|_0^\infty$$

$$= 0 - Q = -Q.$$

In the last step we have assumed that $e^{At} \to 0$ for $t \to \infty$. This is correct only for matrices which have eigenvalues in the left semi-plane (stable). This concludes the proof.

CHAPTER 8
Linear Quadratic Regulators

We have seen how to design a state-space controller by pole location, but the question where to put those desired closed-loop poles is open. Even if the system is controllable and we could choose any location for closed-loop poles, we may get unreasonably high inputs (control effort) depending on the poles' location. In this chapter, we will learn how to design a state controller in a way that the output will converge reasonably fast, and the input will be reasonably small. A good thing about such a design is that we do not need to choose the pole location and there is no change to the connection architecture of the system—it is the same as in Chapter 3.

Cost Function

The system is given by

$$\begin{cases} \dot{x}(t) = Ax(t) + Bu(t) \\ y(t) = Cx(t) \end{cases}$$

$$x(0) = x_0$$

We define a scalar *cost function* J that includes squared functions of control as follows:

$$J = \int_0^\infty \left(\underbrace{x^T(t)Qx(t)}_{\substack{\text{penalizes the transient} \\ \text{state deviation from} \\ \text{the origin}}} + \underbrace{u^T(t)Ru(t)}_{\substack{\text{penalizes the control} \\ \text{effort}}} \right) dt \qquad (8.1)$$

where Q and R are positive definite weight matrices and J is the weighted sum of input and output energies that we want to minimize. Ideally, it would be great to have the integrand decaying to zero immediately, but, in reality, if we preserve small inputs, we cannot rapidly regulate the states and the output. If we push hard to make the states small, then we must use big inputs. There is a need to compromise.

NOTE The expression x^TQx is called a *quadratic form*. It is always scalar (check the dimensions). If we denote the elements of matrix Q by $q_{[ij]}$, then $x^TQx = q_{11}x_1^2 + q_{22}x_2^2 + q_{33}x_3^2 + \cdots + q_{12}x_1x_2 + q_{21}x_2x_1 + \cdots \geq 0$ for positive definite Q.

The goal of the design in this chapter is to minimize J with some controller K, where $u(t) = -Kx(t)$. Such a controller is called an *optimal controller*.

NOTE The same notation works for multiple-input multiple-output (MIMO) systems.

Theorem 8.1 Sufficient and Necessary Conditions for the Existence and Uniqueness of Optimal Controller

The following conditions should be satisfied for the existence and uniqueness of an optimal controller:

1. The pair $\{A, B\}$ is stabilizable.
2. $R \succ 0$ is symmetric.
3. It is possible to write the matrix Q as $0 \leq Q = C_q^T C_q$ (Q is symmetric positive semi-definite).
4. C_q is such a matrix that $\{A, C_q\}$ is a detectable pair.

Continuous-Time Optimal Controller

Theorem 8.2
The optimal controller $u(t) = -Kx(t)$ is given by

$$K = R^{-1}B^T S \tag{8.2}$$

where S is a positive definite matrix, which solves the following *continuous-time algebraic Riccati equation* (CARE):

$$A^T S + SA + Q - SBR^{-1}B^T S = 0 \tag{8.3}$$

In the case of an optimal controller, the optimal cost is

$$J_{min} = x_0^T S x_0 \tag{8.4}$$

NOTE For a single-input single-output (SISO) system, and particular choice of $Q = C^T C$ and $R = \rho$, we get $J = \int_0^\infty (y^2 + \rho u^2)dt; (\rho > 0)$.

Cross-Product Extension of Cost Function

It is possible to generalize the optimization criteria as follows:

$$J = \int_0^\infty (x \ \ u)\begin{pmatrix} Q & N \\ N^T & R \end{pmatrix}\begin{pmatrix} x \\ u \end{pmatrix}dt = \int_0^\infty (x^T Qx + u^T Ru + x^T Nu + u^T N^T x)dt \tag{8.5}$$

Theorem 8.3
The pair $\{A, B\}$ is controllable; $\{A - BR^{-1}N^T, W\}$ is detectable, where W is some matrix for which $WW^T = Q - NR^{-1}N^T$.

The optimal solution is

$$A^T S + SA + Q - (SB + N)R^{-1}(SB + N)^T = 0 \tag{8.6}$$

$$K = R^{-1}(B^T S + N^T) \tag{8.7}$$

Prescribed Degree of Stability

We can extend the cost function in various ways, including increasing the stability margins. The new cost function is

$$J = \int_0^\infty e^{2\alpha t}(u^T Ru + x^T Qx)\, dt \tag{8.8}$$

It is possible to show that in this case the solution is

$$K = R^{-1}B^T S \tag{8.9}$$

where S is the solution of the following Riccati equation:

$$(A + \alpha I)^T S + S(A + \alpha I) + Q - SBR^{-1}B^T S = 0 \tag{8.10}$$

All existence and uniqueness conditions here are the same, but for matrix $A + \alpha I$ instead of matrix A.

Discrete-Time Optimal Controller

The system is given by

$$\begin{cases} x[k+1] = Ax[k] + Bu[k] \\ y[k] = Cx[k] \end{cases}$$

$$x[0] = x_0$$

The cost function is

$$J = \sum_{k=1}^\infty \left[x^T[k]Qx[k] + u^T[k]Ru[k] \right] \tag{8.11}$$

Theorem 8.4
The optimal controller $u(t) = -Kx(t)$ is given by

$$K = (B^T SB + R)^{-1} B^T SA \tag{8.12}$$

where S is the solution of *discrete algebraic Riccati equation* (DARE):

$$A^T SA - S - A^T SB[B^T SB + R]^{-1} B^T SA + Q = 0 \tag{8.13}$$

The optimal cost function is $J_{min} = x_0^T S x_0$.

Solved Problems

Problem 8.1
The system is given by

$$\begin{cases} \dot{x}(t) = Ax(t) + Bu(t) \\ y(t) = Cx(t) \end{cases}$$

$$A = \begin{pmatrix} -2 & 0 \\ 1 & 0 \end{pmatrix}; \quad B = \begin{pmatrix} 1 \\ 1 \end{pmatrix}; \quad C = \begin{pmatrix} 0 & 1 \end{pmatrix}$$

Design the state feedback controller $u(t) = -Kx(t)$ which minimizes the following cost function:

$$J = \int_0^\infty (y^2 + u^2) dt$$

Solution
Formally, we need to check the conditions of Theorem 8.1 to make sure that the solution exists and that it is unique. Technically, we could start with solving the Riccati equation and hope that a positive definite solution for matrix P exists, but that could be a waste of time in cases where the conditions of Theorem 8.1 are not satisfied. In practice, most systems would satisfy all conditions.

Just for the sake of completeness, we will check all four conditions.

1. The pair $\{A, B\}$ is controllable because the controllability matrix $\mathcal{C} = \begin{pmatrix} 1 & -2 \\ 1 & 1 \end{pmatrix}$ is full rank; thus, the system is stabilizable.

2. $R = 1 \succ 0$ is symmetric (as any scalar).

3. We define $C_q = C = (0 \quad 1)$ and can write $Q = C_q^T C_q = \begin{pmatrix} 0 & 0 \\ 0 & 1 \end{pmatrix}$. Then Q is symmetric positive semi-definite (eigenvalues 0 and 1) and $x^T Q x = x^T C^T C x = (Cx)^T (Cx) = y^2$ as in the criteria J given in the question.

4. The observability matrix of the pair $\{A, C_q\}$ is $\mathcal{O} = \begin{pmatrix} 0 & 1 \\ 1 & 0 \end{pmatrix}$, which is full rank. So, C_q is such a matrix that $\{A, C_q\}$ is a detectable pair.

Now, we need to solve the Riccati matrix equation, which is difficult and nonlinear. In the next chapter, we will learn how to avoid solving this equation and getting the same results with symmetric root locus. We substitute A, B, and Q into Riccati Equation (8.3) $A^T S + SA + C^T C - SBR^{-1}B^T S = 0$:

$$\begin{pmatrix} -2 & 1 \\ 0 & 0 \end{pmatrix}\begin{pmatrix} s_1 & s_2 \\ s_2 & s_3 \end{pmatrix} + \begin{pmatrix} s_1 & s_2 \\ s_2 & s_3 \end{pmatrix}\begin{pmatrix} -2 & 0 \\ 1 & 0 \end{pmatrix} + \begin{pmatrix} 0 \\ 1 \end{pmatrix}(0 \quad 1) - \begin{pmatrix} s_1 & s_2 \\ s_2 & s_3 \end{pmatrix}\begin{pmatrix} 1 \\ 1 \end{pmatrix}(1)(1 \quad 1)\begin{pmatrix} s_1 & s_2 \\ s_2 & s_3 \end{pmatrix} = \begin{pmatrix} 0 & 0 \\ 0 & 0 \end{pmatrix}$$

$$\begin{pmatrix} -4s_1 + 2s_2 & -2s_2 + s_3 \\ -2s_2 + s_3 & 0 \end{pmatrix} + \begin{pmatrix} 0 & 0 \\ 0 & 1 \end{pmatrix} - \begin{pmatrix} (s_1 + s_2)^2 & (s_1 + s_2)(s_2 + s_3) \\ (s_1 + s_2)(s_2 + s_3) & (s_2 + s_3)^2 \end{pmatrix} = \begin{pmatrix} 0 & 0 \\ 0 & 0 \end{pmatrix}$$

Linear Quadratic Regulators

On adding these matrices, we get four equations two of which are the same because of symmetricity. Therefore, we need to solve just three equations:

$$\begin{cases} (s_2 + s_3)^2 = 1 \\ (s_1 + s_2)(s_2 + s_3) + 2s_2 - s_3 = 0 \\ (s_1 + s_2)^2 + 4s_1 - 2s_2 = 0 \end{cases}$$

There is no standard way to analytically solve this system of second-degree equations. Generally, these equations are solved numerically, but we will try to make sensible decisions here to solve them by hand. First, equation $(s_2 + s_3)^2 = 1$ could mean $s_2 + s_3 = 1$ or $s_2 + s_3 = -1$. Let's start with $s_2 + s_3 = 1$. Then, $(s_1 + s_2)(s_2 + s_3) + 2s_2 - s_3 = s_1 + s_2 + 2s_2 - s_3 = s_1 + 3s_2 - s_3 = s_1 + 3s_2 - 1 + s_2 = s_1 + 4s_2 - 1 = 0$. Thus, $s_1 = 1 - 4s_2$. We substitute s_1 into the last equation and get $(s_1 + s_2)^2 + 4s_1 - 2s_2 = (1 - 4s_2 + s_2)^2 + 4(1 - 4s_2) - 2s_2 = 9s_2^2 - 24s_2 + 5 = 0$. The solution of that quadratic equation is $s_2 = \frac{4}{3} \pm \frac{\sqrt{11}}{3}$. Substituting this back into $s_1 = 1 - 4s_2$, we get $s_1 = -4\frac{1}{3} \pm \frac{4\sqrt{11}}{3}$. From the Sylvester criterion for positive definite matrices, all minors built on the main diagonal of the matrix S should be positive, including s_1. Thus, the only possible solution is $s_1 = -4\frac{1}{3} + \frac{4\sqrt{11}}{3} = 0.089$; $s_2 = \frac{4}{3} - \frac{\sqrt{11}}{3} = 0.228$; $s_3 = 1 - s_2 = 0.772$. Matrix $S = \begin{pmatrix} 0.089 & 0.228 \\ 0.228 & 0.772 \end{pmatrix}$ is positive definite (can be verified by the Sylvester criterion or direct computation of eigenvalues). Therefore, this P satisfies the condition and we don't need to search for another solution (the solution is unique).

Using Formula (8.2), the controller is $K = R^{-1} B^T S = \frac{1}{1}(1 \ \ 1) \begin{pmatrix} 0.089 & 0.228 \\ 0.228 & 0.772 \end{pmatrix} = (0.317 \ \ 1)$.

Note that we did not need to decide on the location of the closed-loop poles to design this controller.

Nevertheless, we can compute their location where the optimal controller will put them by solving the $\det(sI - A + BK) = 0$ equation (finding eigenvalues of the closed-loop matrix $A - BK$). Those eigenvalues are $-1.67 \pm 0.5j$.

Problem 8.2

The system is given by

$$\dot{x}(t) = Ax(t) + Bu(t)$$
$$y(t) = Cx(t)$$

$$A = \begin{bmatrix} 0 & 1 \\ 0 & 2 \end{bmatrix}; \quad B = \begin{bmatrix} 0 \\ 1 \end{bmatrix}; \quad C = [1 \ \ 0]$$

A. Compute the optimal controller $u(t) = -Kx(t)$ to minimize the following cost function:

$$J = \int_0^\infty (2x_1^2 + 2\sqrt{2} x_1 x_2 + 2x_2^2 + u^2) \, dt$$

Chapter Eight

B. Find the optimal location of closed-loop poles.
C. If the initial state is given by $x(0) = (1 \quad 1)^T$, find J_{min}.

Solution

A. First, we need to compute matrices Q and R from the cost function J. From the SISO case, we have $x^T Q x + u^T R u = x^T \begin{pmatrix} q_1 & q_2 \\ q_2 & q_3 \end{pmatrix} x + \rho u^2 = q_1 x_1^2 + 2 q_2 x_1 x_2 + q_3 x_2^2 + \rho u^2$.

From the comparison of coefficients between the last expression and the integrand of J, we get $R = 1$ and $Q = \begin{pmatrix} 2 & \sqrt{2} \\ \sqrt{2} & 2 \end{pmatrix}$. It is easy to see that $Q \succ 0$ (Sylvester criterion). So, we need to solve Riccati Equation (8.3) and then compute the vector of controller's gains using (8.2).

$$A^T S + S A + Q - S B B^T S = 0$$

$$\begin{pmatrix} 0 & 0 \\ 1 & 2 \end{pmatrix} \begin{pmatrix} s_1 & s_2 \\ s_2 & s_3 \end{pmatrix} + \begin{pmatrix} s_1 & s_2 \\ s_2 & s_3 \end{pmatrix} \begin{pmatrix} 0 & 1 \\ 0 & 2 \end{pmatrix} + \begin{pmatrix} 2 & \sqrt{2} \\ \sqrt{2} & 2 \end{pmatrix} - \begin{pmatrix} s_1 & s_2 \\ s_2 & s_3 \end{pmatrix} \begin{pmatrix} 0 \\ 1 \end{pmatrix} (0 \quad 1) \begin{pmatrix} s_1 & s_2 \\ s_2 & s_3 \end{pmatrix} = \begin{pmatrix} 0 & 0 \\ 0 & 0 \end{pmatrix}$$

After multiplying and adding the matrices above, we get three quadratic equations with three unknowns:

$$\begin{cases} 2 - s_2^2 = 0 \\ s_1 + 2 s_2 + \sqrt{2} - s_2 s_3 = 0 \\ 2(s_2 + 2 s_3) + 2 - s_3^2 = 0 \end{cases}$$

From the first equation, $s_2 = \pm\sqrt{2}$. If we choose $s_2 = -\sqrt{2}$, then the third equation becomes $2(-\sqrt{2} + 2 s_3) + 2 - s_3^2 = 0$, which has solutions $s_3 = 2 \pm \sqrt{6 - 2\sqrt{2}} =$ 3.78 or 0.22. On substituting the numbers into the second equation, we get $s_1 - 2\sqrt{2} + \sqrt{2} + \sqrt{2} \cdot 3.78 = 0$ or $s_1 - 2\sqrt{2} + \sqrt{2} + \sqrt{2} \cdot 0.22 = 0$. Therefore, $s_1 = -3.93$ or $s_1 = 1.103$. From the Sylvester criterion, only positive s_1 will work. The conclusion is that $S = \begin{pmatrix} 1.103 & -\sqrt{2} \\ -\sqrt{2} & 0.22 \end{pmatrix}$, but this matrix is not positive definite; thus, the choice of $s_2 = -\sqrt{2}$ was incorrect. Now, let's choose $s_2 = \sqrt{2}$ and follow the same steps to get positive definite matrix $S = \begin{pmatrix} 2.79 & \sqrt{2} \\ \sqrt{2} & 4.97 \end{pmatrix}$. We compute the controller using this matrix:

$$K = B^T S = (\sqrt{2} \quad 4.97)$$

B. To find optimal locations of closed-loop poles, we find eigenvalues of the closed-loop matrix $A - BK$ with the vector of gains K computed in the previous part.

$$\det(sI - A + BK) = \det\left(\begin{pmatrix} s & 0 \\ 0 & s \end{pmatrix} - \begin{pmatrix} 0 & 1 \\ 0 & 2 \end{pmatrix} + \begin{pmatrix} 0 \\ 1 \end{pmatrix}(\sqrt{2} \quad 4.97)\right)$$

$$= \det\begin{pmatrix} s & -1 \\ \sqrt{2} & s + 2.97 \end{pmatrix} = s(s + 2.97) + \sqrt{2} = s^2 + 2.97 s + \sqrt{2} = 0$$

The solutions of that quadratic equation are −2.38 and −0.595. Obviously, both closed-loop poles are stable in the left semi-plane. This will always be true for optimal controllers.

C. Based on (8.4), $J_{min} = x_0^T S x_0 = 10.59$.

Problem 8.3
Derive the continuous-time algebraic Riccati equation.

Solution
Optimal controller $u = -Kx$ minimizes the cost function $J = \int_0^\infty x^T Q x + u^T R u \, dt$. If we substitute u into the integral, $J = \int_0^\infty x^T Q x + (-Kx)^T R(-Kx) dt = \int_0^\infty x^T Q x + x^T K^T R K x \, dt = \int_0^\infty x^T (Q + K^T R K) x \, dt$.

The closed-loop system $\dot{x} = (A - BK)x; x(0) = x_0$ has a general solution [Formula (2.9)] $x(t) = e^{(A-BK)t} x_0$, thus $J = \int_0^\infty (e^{(A-BK)t} x_0)^T (Q + K^T R K) e^{(A-BK)t} x_0 \, dt = \int_0^\infty x_0^T e^{(A-BK)^T t} (Q + K^T R K) e^{(A-BK)t} \times x_0 \, dt = x_0^T \left(\int_0^\infty e^{(A-BK)^T t} (Q + K^T R K) e^{(A-BK)t} \, dt \right) x_0$.

Let's define a matrix $S = \int_0^\infty e^{(A-BK)^T t} (Q + K^T R K) e^{(A-BK)t} \, dt$. This matrix is positive definite (by definition) if $J = x_0^T S x_0 > 0$.

In Problem 7.5 of Chapter 7 we have shown that $P = \int_0^\infty e^{A^T t} Q e^{At} \, dt$ solves the Lyapunov equation $A^T P + PA = -Q$ for stable A. This means that S solves the following equation:

$$(A - BK)^T S + S(A - BK) = -(Q + K^T R K)$$

where we just replaced A with $A - BK$ and Q with $Q + K^T R K$.

After opening parentheses

$$A^T S - K^T B^T S + SA - SBK + Q + K^T R K = 0$$

Completing the square

$$A^T S + SA + Q + (SBR^{-1} - K^T) R (R^{-1} B^T S - K) - SBR^{-1} B^T S = 0$$

The equation above is satisfied if we choose $K = R^{-1} B^T S$ and S solves $A^T S + SA + Q - SBR^{-1} B^T S = 0$, which is the algebraic Riccati equation we need.

Problem 8.4
The system is given by

$$\begin{cases} \dot{x}(t) = Ax(t) + Bu(t) \\ y(t) = Cx(t) \end{cases}, \quad x(0) = (\alpha \quad 0)^T$$

$$A = \begin{pmatrix} 0 & 0 \\ 0 & 1 \end{pmatrix} \quad B = \begin{pmatrix} 1 \\ 1 \end{pmatrix} \quad C = (1 \quad 0)$$

Chapter Eight

We would like to design an optimal controller $u(t) = -Kx(t)$ that minimizes the following optimality criterion (cost function): $J = \int_0^\infty (y^2 + u^2)dt$.

A. Find *all* three solutions of the Riccati equation.

B. Show that only two of those solutions are positive semi-definite $S \geq 0$; and thus, only they can be candidates for the solution. Compute the cost function J for both. Which solution is better? Would your answer be different if $x(0) = (\alpha \quad \beta)^T$?

C. For each solution from (B), compute the controller K, and the appropriate closed-loop eigenvalues. Which design is better?

D. Explain the difference between (B) and (C).

Solution

A. The Riccati equation: $A^TS + SA + C^TC - SBB^TS = 0$

$$\begin{pmatrix} 0 & 0 \\ 0 & 1 \end{pmatrix}\begin{pmatrix} s_1 & s_2 \\ s_2 & s_3 \end{pmatrix} + \begin{pmatrix} s_1 & s_2 \\ s_2 & s_3 \end{pmatrix}\begin{pmatrix} 0 & 0 \\ 0 & 1 \end{pmatrix} + \begin{pmatrix} 1 & 0 \\ 0 & 0 \end{pmatrix} - \begin{pmatrix} s_1 & s_2 \\ s_2 & s_3 \end{pmatrix}\begin{pmatrix} 1 \\ 1 \end{pmatrix}(1 \quad 1)\begin{pmatrix} s_1 & s_2 \\ s_2 & s_3 \end{pmatrix} = 0$$

$$\begin{pmatrix} 0 & s_2 \\ s_2 & 2s_3 \end{pmatrix} + \begin{pmatrix} 1 & 0 \\ 0 & 0 \end{pmatrix} - \begin{pmatrix} (s_1 + s_2)^2 & (s_1 + s_2)(s_2 + s_3) \\ (s_1 + s_2)(s_2 + s_3) & (s_2 + s_3)^2 \end{pmatrix} = 0$$

Thus, $(s_1 + s_2)^2 = 1$; $(s_1 + s_2)(s_2 + s_3) = s_2$; $2s_3 = (s_2 + s_3)^2$.

Case 1: $s_1 = 1$, $s_2 = s_3 = 0$, thus $S_1 = \begin{pmatrix} 1 & 0 \\ 0 & 0 \end{pmatrix} \succcurlyeq 0$.

Case 2: $s_1 = -1$, $s_2 = s_3 = 0$, thus $S_2 = \begin{pmatrix} -1 & 0 \\ 0 & 0 \end{pmatrix}$ not positive semi-definite.

Case 3: $s_1 = 3$, $s_2 = -4$, $s_3 = 8$, thus $S_3 = \begin{pmatrix} 3 & -4 \\ -4 & 8 \end{pmatrix} \succcurlyeq 0$.

B. S_1 and S_3 are candidates for the solution. For the initial state $(\alpha \quad 0)^T$, we get

$$J_1 = x_0^T S x_0 = \alpha^2; \qquad J_3 = 3\alpha^2$$

Thus, $J_1 \leq J_3$, and we choose S_1.
For the initial state $(\alpha \quad \beta)^T$, we get

$$J_1 = \alpha^2; \qquad J_3 = 3\alpha^2 - 8\alpha\beta + 4\beta^2$$

But $J_3 - J_1 = 2(\alpha - 2\beta)^2 \geq 0$, thus, $J_1 \leq J_3$, and we choose S_1 again.

C. $K_1 = B^T S_1 = (1 \quad 0) \rightarrow A - BK_1 = \begin{pmatrix} -1 & 0 \\ -1 & 1 \end{pmatrix} \rightarrow \lambda = \pm 1$

$K_3 = B^T S_3 = (1 \quad 0) \rightarrow A - BK_3 = \begin{pmatrix} 1 & -4 \\ 1 & -3 \end{pmatrix} \rightarrow \lambda = -1, -1$

Thus, for stability we have to choose S_3, and that is the only possible solution.

D. The reason for the minimizing solution which is not stabilizing the system is the absence of observability. The eigenvalue that is not observable is not stable. Thus, the criterion does not depend on everything that happens to the system and can be minimal even though part of the system is diverging.

CHAPTER 9
Symmetric Root Locus

In Chapter 8, we discussed the designing of an optimal controller, but we didn't explicitly answer the question about where the optimal poles should be. Of course, it is possible to compute the eigenvalues of $A - BK_{opt}$, but if we want to know how the closed-loop poles move if we change the cost function weights, we need symmetric root locus (SRL). As a by-product, in many simple cases we get the optimal location of the closed-loop poles without solving the Riccati equations. After knowing the location of the poles, we should be able to design an optimal controller using the standard pole placement techniques from Chapter 3.

Continuous-Time SRL

The system is given by

$$\begin{cases} \dot{x}(t) = Ax(t) + Bu(t) \\ y(t) = Cx(t) \end{cases} ; \quad x(0) = x_0$$

Theorem 9.1 (Letov, 1960)

For a minimal linear time invariant (LTI) single-input single-output (SISO) system with transfer function $G(s) = b(s)/a(s)$, the optimal eigenvalues that minimize $J = \int_0^\infty (y^2 + \rho u^2) dt$ are given by the roots with negative real part of the following equation:

$$a(s)a(-s) + \frac{1}{\rho} b(s)b(-s) = 0 \tag{9.1}$$

NOTES

1. The total number of roots for this equation is twice the system's order.
2. The cost function given here is a simple case of the cost function given in the previous chapter. Extensions for more sophisticated cases exist.

Using Letov's Theorem 9.1, we can design the optimal controller as follows:

1. Solve Equation (9.1) and find $2n$ roots.
2. Choose n "stable" roots and create the desired closed-loop polynomial $\alpha(s)$ from them.
3. Compute the controller using $\alpha(s)$ and $a(s)$ (Bass-Gura, Ackermann, or comparison of coefficients).

Note that Equation (9.1) is symmetrical, that is, if we exchange s with $-s$, we will get the same equation. In other words, if s_i is the root of (9.1), then $-s_i$ is also the root of (9.1) and all the roots have their mirror image roots with regard to a vertical axis. Since the solutions of Equation (9.1) include all optimal poles [solutions of $\alpha(s) = 0$], it will also include their mirror images [solutions of $\alpha(-s) = 0$], where α is the desired closed-loop characteristic polynomial.

Thus, in the closed loop we have

$$\alpha(s)\alpha(-s) = a(s)a(-s) + \frac{1}{\rho}b(s)b(-s) = 0 \qquad (9.2)$$

It is still not very clear what the relation of (9.1) with the root locus is. Here is an explanation. When dividing both sides of (9.1) by $a(s)a(-s)$, we get

$$1 + \frac{1}{\rho}\frac{b(s)b(-s)}{a(s)a(-s)} = 1 + \frac{1}{\rho}G(s)G(-s) = 0$$

If we think about $1/\rho$ as gain and the total open-loop transfer function $H(s) = G(s)G(-s)$, this is exactly the format we need to plot the standard *root locus (RL)* of W. R. Evans:

$$(-1)^{m-n}\frac{1}{\rho}\frac{\prod(s+z_i)(s-z_i)}{\prod(s+p_i)(s-p_i)} = -1 \qquad (9.3)$$

where z_i are the roots of $b(s)$ and p_i are the roots of $a(s)$, and $\deg(a(s)) = m$, $\deg(b(s)) = n$.

NOTES

1. We need to choose RL gain $= 1/\rho$ (unfortunately for our notation this gain is also denoted by K in the standard RL).
2. The obtained RL is symmetric with respect to $s = j\omega$ axis.

To plot the *symmetric RL* (SRL) as a function of $1/\rho$, we need to plot open-loop poles and zeros of $G(s)$. Then, we mirror all poles and zeros with respect to the imaginary axis. Finally, we plot a regular RL based on all these poles and zeros.

The last question is what sign we should pick for the gain (remember that there is one root locus for a negative and one for a positive gain). From the definition, we see that for even $m-n$ the gain should be positive, and for odd $m-n$ the gain should be negative.

CAUTION! SRL never crosses the $j\omega$ axis! (It may start on the axis though.)

Discrete-Time SRL

The system is given by

$$\begin{cases} x[k+1] = Ax[k] + Bu[k] \\ y[k] = Cx[k] \\ x[0] = x_0 \end{cases}$$

Theorem 9.2 (Letov, 1960)

For a minimal LTI SISO system with transfer function $G(z) = b(z)/a(z)$, the optimal eigenvalues that minimize $J = \sum_{k=0}^{\infty}(y^2[k] + \rho u^2[k]); \rho > 0$ are given by the roots with negative real part of the following equation:

$$a(z)a(z^{-1}) + \frac{1}{\rho}b(z)b(z^{-1}) = 0 \tag{9.4}$$

Similarly, to continuous-time case, we create the desired polynomial $\alpha(z)$ from "stable" roots of the equation above (in the unit circle).

The SRL is produced by using the poles and zeros of $G(z)$ and $G(z^{-1})$. Hence the name SRL is less appropriate, since the RL graph is not symmetric anymore.

CAUTION! SRL never crosses the unit circle!

How to Sketch Continuous-Time SRL

Modern software could solve most of the problems described in this book in no time. Plotting the root locus using software is easy as well. Sometimes, it is still useful to sketch a root locus by hand to understand how things work. In most cases, it is not essential to compute splitting points or angles of asymptotes precisely. This leaves only a few basic rules to remember that will not require any computations. They include:

1. Draw a system of coordinates with real and imaginary axes. Use the open-loop transfer function $G(s)$ and draw its zeros and poles in the complex domain (system of coordinates).

2. Mirror all poles and zeros about the imaginary axis [in total you should have twice the number of zeros and poles as compared to the original transfer function $G(s)$].

3. Identify intervals on the real axis where the poles will be moving. The intervals are identified by counting the total number of poles and zeros to the right from the point of interest in the interval. If the total number to the right is even (odd) then there is a motion on that interval. The tricky part here is to decide if we are checking an even number or an odd number. Obviously, they exclude each other. One will include interval crossing the origin and another will not. So, we must choose an SRL that does not cross the origin.

4. All poles should end up at zero or approach infinity when $1/\rho$ approaches infinity.
5. All poles are moving symmetrically about the real and imaginary axes.

These steps and rules are enough to plot the correct SRL.

How to Sketch Discrete-Time SRL

Discrete-time SRL plotting is very similar to continuous time. Following are the required rules to remember:

1. Draw a system of coordinates with real and imaginary axes. Use the open-loop transfer function $G(z)$ and draw its zeros and poles in the complex domain (system of coordinates).
2. Mirror all poles and zeros about the unit circle by plotting the reciprocal of all poles and zeros [in total you should have twice the number of zeros and poles as compared to the original transfer function $G(z)$].
3. Identify intervals on the real axis where the poles will be moving. The intervals are identified by counting the total number of poles and zeros to the right from the point of interest in the interval. If the total number to the right is even (odd), then there is a motion on that interval. The tricky part here is to decide if we are checking an even number or an odd number. Obviously, they exclude each other. So, we must choose an SRL that does not cross the unit circle.
4. All poles should end up at zero or approach infinity when $1/\rho$ approaches infinity.
5. All poles are moving symmetrically about the real axis.

Solved Problems

Problem 9.1
The system is given by

$$\dot{x}(t) = Ax(t) + Bu(t)$$
$$y(t) = Cx(t)$$

$$A = \begin{pmatrix} -2 & 0 \\ 1 & 0 \end{pmatrix}; \quad B = \begin{pmatrix} 1 \\ 1 \end{pmatrix}; \quad C = (0 \quad 1)$$

Design the state feedback controller $u(t) = -Kx(t)$, which minimizes the following cost function:

$$J = \int_0^\infty (y^2 + u^2) dt$$

Solution

The open-loop transfer function is computed using Formula (2.6):

$$G(s) = \frac{b(s)}{a(s)} = C(sI - A)^{-1}B + D = (0 \ \ 1)\begin{pmatrix} s+2 & 0 \\ -1 & s \end{pmatrix}^{-1}\begin{pmatrix} 1 \\ 1 \end{pmatrix} + 0$$

$$= (0 \ \ 1)\frac{1}{s(s+2)}\begin{pmatrix} s & 0 \\ 1 & s+2 \end{pmatrix}\begin{pmatrix} 1 \\ 1 \end{pmatrix} = \frac{1}{s(s+2)}(1 \ \ s+2)\begin{pmatrix} 1 \\ 1 \end{pmatrix} = \frac{s+3}{s(s+2)}$$

Thus, $b(s) = s+3$ and $a(s) = s(s+2)$. From the definition of cost function J, we can identify the multiplier of u^2 as $\rho = 1$.

Now, using Letov's Equation (9.1):

$$a(s)a(-s) + \frac{1}{\rho}b(s)b(-s) = s(s+2)(-s)(-s+2) + (s+3)(-s+3) = 0$$

$$s^2(s^2-4) - (s^2-9) = s^4 - 5s^2 + 9 = 0$$

This is a biquadratic equation that first solved for s^2:

$$s^2 = \frac{5 \pm \sqrt{25-36}}{2} = 2.5 \pm \frac{\sqrt{11}}{2}j$$

By computing the complex square root using de Moivre's formula, we get four roots $s_{1,2,3,4} = \pm 1.658 \pm 0.5j$. We pick only stable roots $s_{1,2} = -1.658 \pm 0.5j$.

Though not required in this problem, it is often useful to sketch the SRL to make sure that the computed poles do not lie outside the SRL. Figure 9.1 shows the plot of SRL and the location of the optimal pole for $\rho = 1$.

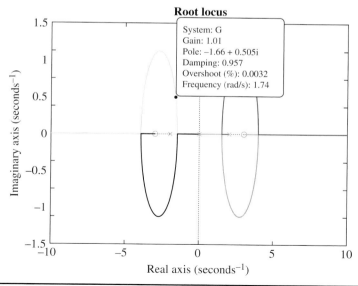

Figure 9.1 Symmetric root locus for Problem 9.1.

Now, we need to use one of the methods discussed in Chapter 3 to design a controller. For example, we could compare the coefficients of the desired polynomial $\alpha(s)$ and the closed-loop polynomial $a_K(s)$:

$$\begin{cases} \alpha(s) = (s+1.658-0.5j)(s+1.658+0.5j) = s^2 + 3.317s + 3 \\ a_K(s) = \det(sI - A + BK) = \det\begin{pmatrix} s+2+k_1 & k_2 \\ -1+k_1 & s+k_2 \end{pmatrix} = (s+2+k_1)(s+k_2) - k_2(k_1-1) \\ \qquad\qquad = s^2 + (2+k_1+k_2)s + 3k_2 \end{cases}$$

Both polynomials should be the same; thus, $s^2 + (2+k_1+k_2)s + 3k_2 = s^2 + 3.317s + 3$ and from comparison of the coefficients:

$$\begin{cases} 2 + k_1 + k_2 = 3.317 \\ 3k_2 = 3 \end{cases}$$

Therefore, $k_1 = 0.317$ and $k_2 = 1$, $K = (0.317 \quad 1)$. This is the same result we obtained for Problem 8.1 in Chapter 8 for the same system.

Problem 9.2

The system is given in a canonical controller form, and its transfer function is

$$G(s) = \frac{s-1}{(s-1)(s+3)}$$

A. Is this system detectable?

We define the following optimality criterion: $J = \int_0^\infty \left(x_1^2 + 9x_2^2 + 6x_1 x_2 + \frac{1}{3} u^2 \right) dt$.

B. Draw the SRL of the system. What is the optimal poles position?
C. Describe how you would design the state feedback controller to move all poles to the optimal locations of part (B).

Solution

A. The system is not minimal because of the cancelation of $s-1$. Also, the system is controllable because it is given in the canonical controller form. This means that the system must not be observable (based on Theorem 2.5, observable and controllable system must be minimal). Since the canceled value $s = 1$ is unstable, the system is not detectable.

B. From the definition of cost function J, we identify $\rho = \frac{1}{3}$. Unfortunately, the rest of the cost function is not given by y^2 as in Theorem 9.1.

Here we have to use one important trick. Our goal is to bring the cost function to the standard form to be able to use Letov's equation. Note that the system's output is not at all used in the optimal (or state space in general) controller design. In other words, the system's matrix C is not participating in any computations. So, for any C we are supposed to get the same optimal controller. What if we replace the original matrix C with $\tilde{C}$ such that $(\tilde{C}x)^T(\tilde{C}x) = \tilde{y}^2 = x_1^2 + 9x_2^2 + 6x_1 x_2$? This way we get a different output $\tilde{y}$, but this does not matter for the computation of the optimal controller. Such $\tilde{C}$ exists and you could easily verify that $\tilde{C} = (1 \ \ 3)$.

The original system's implementation in canonical controller form is given by

$$A = \begin{pmatrix} -2 & 3 \\ 1 & 0 \end{pmatrix}; \quad B = \begin{pmatrix} 1 \\ 0 \end{pmatrix}; \quad C = (1 \ \ -1)$$

Now, we replace the C with $\tilde{C}$ to find the transfer function of interest $\tilde{G}(s)$ for which we minimize $J = \int_0^\infty \left(\tilde{y}^2 + \frac{1}{3} u^2 \right) dt$:

$$\tilde{G}(s) = \frac{b(s)}{a(s)} = \tilde{C}(sI - A)^{-1} B = \frac{s+3}{s^2 + 2s - 3} = \frac{1}{s-1}$$

Thus, $b(s) = 1$ and $a(s) = s - 1$.

The SRL equation is

$$a(s)a(-s) + 3 = 0$$
$$(s-1)(-s-1) + 3 = 0$$
$$-s^2 + 1 + 3 = 0$$
$$s_{1,2} = \pm 2$$

Note that we have got only one stable pole, while the system is of the second order; thus, we need two poles. One of the eigenvalues was canceled ($s = -3$), so we cannot move it with the controller and must design the controller to leave this pole as in the closed loop. The SRL plot is given in Figure 9.2.

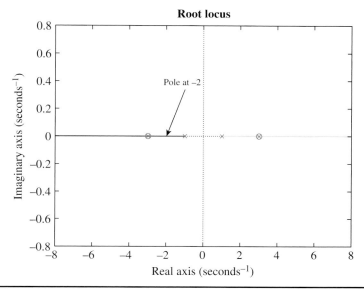

FIGURE 9.2 Symmetric root locus for Problem 9.2.

C. We chose the optimal location of poles at −3 and −2 based on the solution in the previous part of the problem, and use the Bass-Gura, Ackermann, or comparison of coefficients techniques to design the controller for the original canonical controller implementation.

Problem 9.3
The system is given by

$$\dot{x}(t) = \begin{pmatrix} 0 & 1 \\ -2 & 0 \end{pmatrix} x(t) + \begin{pmatrix} 0 \\ 1 \end{pmatrix} u(t)$$
$$y(t) = (2 \quad 0) x(t)$$

with the cost function $J = \int_0^\infty (y^2 + \rho u^2) dt$.

A. Find the open-loop transfer function: $G(s) = \dfrac{b(s)}{a(s)}$.

B. Draw the SRL and compute the values of closed-loop eigenvalues for $\rho = 1$. Find the appropriate controller K.

C. Find the controller directly by solving the Riccati equation.

Solution

A. The transfer function is $G(s) = C(sI - A)^{-1} B = \dfrac{2}{s^2 + 2}$.

B. From (A), $G(s)G(-s) = \dfrac{4}{(s^2+2)^2}$; thus, Letov's equation is $(s^2+2)^2 = -4 \Rightarrow s^2 + 2 = \pm 2j \rightarrow s^2 = -2 \pm 2j$.

The appropriate SRL is shown in Figure 9.3. Note that it starts and splits from the imaginary axis.

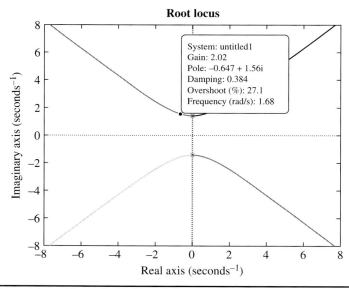

Figure 9.3 Symmetric root locus for Problem 9.3.

The complex quadratic equations above have four roots, but only two of them are stable, $s_{1,2} = -0.643 \pm 1.55j$. The desired closed-loop polynomial is $\alpha(s) = (s + 0.643 + 1.55j)(s + 0.643 - 1.55j) = s^2 + 1.287s + 2.828$. The open-loop polynomial is $a(s) = s^2 + 2$. Now we can apply the Bass-Gura formula:

$$K = (\alpha - a)\mathcal{C}_c \mathcal{C}^{-1} = [(1.287 \quad 2.828) - (0 \quad 2)] \begin{pmatrix} 1 & 0 \\ 0 & 1 \end{pmatrix} \begin{pmatrix} 0 & 1 \\ 1 & 0 \end{pmatrix}^{-1} = (0.828 \quad 1.287)$$

C. By solving the Riccati equation, we obtain $P = \begin{pmatrix} 3.64 & 0.828 \\ 0.828 & 1.2871 \end{pmatrix} \succ 0 \rightarrow K = B^T P = (0.828 \quad 1.287)$.

CHAPTER 10
Kalman Filter

We have seen that the process of state controller design is actually a pole placement process. We have also converted the problem of choosing the desired pole locations into the problem of cost function optimization. Now, we should ask the same question with regard to the observer's poles. Where should we choose them? This is a dual problem, and has a dual solution solved by linear optimal observer, which is a causal version of the Wiener filter. It was proposed by Rudolf Kalman and Richard Bucy (Kalman, 1960; Kalman and Bucy, 1961) even before the "official" development of the state observers by David Luenberger (1964). If we go further back in history, Russian physicist, Ruslan Stratonovich, published about more general nonlinear filters (Stratonovich, 1959). Since the Kalman-Bucy filter is a special case of the Stratonovich filter, it is sometimes called the Stratonovich-Kalman-Bucy filter. The idea is that the optimal observer should minimize the mean squared error between the real state and the estimated one. Since the 1960s, this idea has grown into a broad theory of system estimation and has been widely used in many engineering domains ranging from computer vision and stock market prediction to self-driving cars.

The Idea of Optimal Observer (Estimator) in Presence of Noise

Let's start with a simple example. We have a system of a free fall. A hand releases a ball to fall freely under the influence of gravity and a distance (depth) camera measures the distance to the ball from the initial release point (Figure 10.1).

The ball starts to fall at time t_0. The initial location of the ball is known but measured with some tolerance (imprecision). For the Kalman filter we will always assume that this tolerance is given by a normal (Gaussian) distribution (see Appendix H.3). In other words, most probably the ball is where we measured it to be, but it is also probable that it is somewhere nearby, and much less likely that it is far away from the measured point. The likelihood of the ball being at a specific location is outlined in Figure 10.2.

Assume that we developed a state-space dynamic discrete-time model of this system. We could predict where the ball will be at the next time point $t_1(k = 1)$ using the model alone without any measurements. It is easy to understand from a physics perspective that with constant acceleration the ball will pass longer and longer distances per the same time interval. As time passes by, disturbances and noise are added to the system; thus, our certainty about the predicted location should be lower. It is modeled by widening the Gaussian bell-shape curve (see Figure 10.3) showing our uncertainty about the real location.

Now, we measure the location using the depth camera and get some specific distance (see Figure 10.4). Obviously, the measurement is not perfect, so we still have uncertainty about the real location. This uncertainty in measurement or depth camera tolerance is also modeled as Gaussian distribution given in Figure 10.4.

106 Chapter Ten

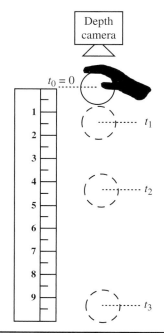

FIGURE 10.1 Hand releasing a ball into a free fall.

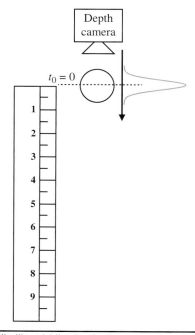

FIGURE 10.2 Initial location likelihood (distribution).

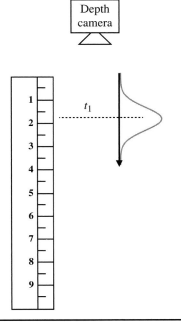

FIGURE 10.3 Predicted ball location distribution.

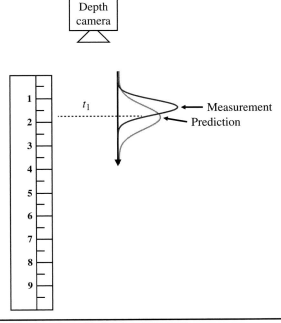

FIGURE 10.4 Updating the location with the measurement.

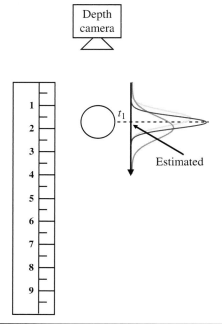

FIGURE 10.5 Final estimated location.

Note that in most cases the predicted location disagrees with the measured location. What should we pick then? Should we trust the model (prediction) or the measurement? We will trust both, but in a certain proportion. The Kalman filter result will be a weighted average of the prediction and the measurement, somewhere in between the two Gaussian peaks. The wider the Gaussian peak, the more uncertainty we have; thus, its weight (relative importance) should be lower. In other words, if, say, we trust the measurement over the model, then the estimated location should be closer to the measurement. It is achieved by multiplying two distributions and getting a final estimated location distribution (generally narrower than each of the two distributions), as shown in Figure 10.5.

NOTE Multiplication of two Gaussian functions is also a Gaussian function.

The following section formalizes the Kalman filter approach.

Optimal Observer (Kalman Filter)

The system is given by

$$\begin{cases} x[k+1] = Ax[k] + Bu[k] + Gw[k] \\ y[k] = Cx[k] + v[k] \\ x[0] \sim N(\bar{x}[0], p[0]) \end{cases} \quad (10.1)$$

where $w[k]$ represents the *system's noise,* and $v[k]$ represents the *measurement noise* (sensor's noise).

We assume that both noises are Gaussian (normally distributed), stationary, and have zero mean:

$$E(w[k]) = 0 \quad E(v[k]) = 0$$
$$E(w[k]w^T[n]) = W\delta[k-n]$$
$$E(v[k]v^T[n]) = V\delta[k-n] \quad (10.2)$$
$$E(w[k]v^T[n]) = 0 : \forall k, n$$

A low w means that you should trust the model; a low v indicates that you should trust the sensors.

The problem is to estimate $x[k]$ from noisy measurements to minimize mean squared error:

$$e[k] = x[k] - \hat{x}[k]$$
$$P[k] = E(e[k]e^T[k]) \quad (10.3)$$
$$J = E(e^T[k]e[k]) = \text{error variance} \rightarrow \min$$

The solution to this problem is the *Kalman filter*.

Theorem 10.1
The condition for minimal variance asymptotically stable filter is as follows:

The pair $\{A, C\}$ is detectable, V is positive definite, and there exists matrix H such that $HH^T = W$ and pair $\{A, GH\}$ is stabilizable.

If this condition is true, then the optimal observer is given by

$$L = PC^T(CPC^T + V)^{-1}$$
$$\hat{x}[k] = (I - LC)(A\hat{x}[k-1] + Bu[k-1]) + Ly[k] \quad (10.4)$$

where P is the solution of the following *algebraic Riccati equation* (ARE):

$$APA^T + GWG^T - APC^T(CPC^T + V)^{-1}CPA^T = P \quad (10.5)$$

Alternatively,

$$L = PC^TV^{-1} \quad (10.6)$$
$$P = M - MC^T[CMC^T + V]^{-1}CM$$
$$M = APA^T + GWG^T \quad (10.7)$$

NOTE The solution of this Riccati equation is different from the previous one, but it produces the same observer's gain L.

Kalman has proposed the recursive solution. In this solution, matrix P and the estimated state are obtained only using the previously computed data in time.

Recursive Solution

The before and after measurement values are denoted by superscripted "−" or "+", respectively. To recursively compute the estimated state $\hat{x}^{\pm}[k]$, we assume $y[-1] = 0$ (for the time $k = 0^-$), and a priori estimations:

$$\hat{x}^-[0] = E(x[0]|y[-1]) = \bar{x}[0]$$
$$P^-[0] = COV(x[0]|y[-1]) = p[0] \tag{10.8}$$

Now we need *measurement update* ($k \geq 0$):

$$\left.\begin{array}{r}P^-[k] \\ \hat{x}^-[k]\end{array}\right\} \Rightarrow \begin{cases}P^+[k] \\ \hat{x}^+[k]\end{cases}$$

$$K[k] = P^-[k]C^T(CP^-[k]C^T + V)^{-1}$$
$$\hat{x}^+[k] = \hat{x}^-[k] + K[k](y[k] - C\hat{x}^-[k]) \tag{10.9}$$
$$P^+[k] = P^-[k] - K[k]CP^-[k]$$

and *time update*:

$$\left.\begin{array}{r}P^+[k] \\ \hat{x}^+[k]\end{array}\right\} \Rightarrow \begin{cases}P^-[k+1] \\ \hat{x}^-[k+1]\end{cases}$$

$$L[k] = AK[k] = AP^-[k]C^T(CP^-[k]C^T + V)^{-1}$$
$$\hat{x}^-[k+1] = A\hat{x}^+[k] + Bu[k] = A\hat{x}^-[k] + Bu[k] + L[k](y[k] - C\hat{x}^-[k]) \tag{10.10}$$
$$P^-[k+1] = AP^+[k]A^T + GWG^T = A(I - K[k]C)P^-[k]A^T + GWG^T$$

For the following time steps $k > 0$, the computations (10.9) and (10.10) are repeated in a loop.

NOTE Make sure that you understand how to obtain all variables and expressions on the right side of the Kalman equations (10.9) and (10.10) given the model matrices, noise parameters, and previous time steps. It is especially important to understand that $k+1$ from the time update step becomes k when $k \leftarrow k+1$ and the computation loops to the measurement update.

The *steady-state covariance matrix* is

$$P_\infty = \lim_{k \to \infty}(P^+[k]) \text{ (in steady state } P^+[k] = P^+[k-1]) \tag{10.11}$$

The *steady-state gain* is

$$K_\infty = \lim_{k \to \infty}(K[k]), \quad L_\infty = AK_\infty \qquad (10.12)$$

Theorem 10.2
The steady-state Kalman filter is stable if and only if all the eigenvalues of $(A - K_\infty CA)$ are inside the unit circle.

NOTES

1. For non-Gaussian noises, the Kalman filter algorithm implements linear optimal filter.
2. The algorithm also works for time varying systems $(A[k], B[k], C[k], V[k], W[k])$.
3. Matrix P can be computed offline if $(A[k], B[k], C[k], V[k], W[k])$ are known in advance.
4. For linear time invariant (LTI) systems, we still get $L[k]$ dependent on time, since the initial conditions do not fit the steady state. It is possible to show (with some restrictions) that $P^-[k] \xrightarrow{k \to \infty} P^-_\infty$, $L[k] \xrightarrow{k \to \infty} L$. The filter designed with this L is called the *stationary Kalman filter*.
5. In some books and articles, matrices W and V are denoted by Q_w and R_v, respectively.

Alternative Kalman Filter Formulation for Unknown Initial Conditions

Theoretically, if there is no initial information available on the estimated process, we should choose $P^-[0] = \infty$, and in that case the equations will not work. Practically, Formulas (10.9) and (10.10) will work when we have high uncertainty about initial conditions. We just need to pick very high values for $P^-[0]$, but the convergence to real state could be slow. We need alternative representation with inversed covariance matrices.

The a priori estimations would be given by $(P^-[0])^{-1}$ and $\hat{x}^-[0]$, where $(P^-[0])^{-1}$ could be zero.

Measurement update ($k \geq 0$):

$$\begin{aligned} K[k] &= P^-[k]C^T(CP^-[k]C^T + V)^{-1} \\ \hat{x}^+[k] &= \hat{x}^-[k] + K[k](y[k] - C\hat{x}^-[k]) \\ (P^+[k])^{-1} &= (P^-[k])^{-1} + C^T V^{-1} C \end{aligned} \qquad (10.13)$$

Time update:

$$\begin{aligned} \hat{x}^-[k+1] &= A\hat{x}^+[k] + Bu[k] \\ P^-[k+1] &= AP^+[k]A^T + GWG^T \end{aligned} \qquad (10.14)$$

The block diagram in Figure 10.6 explains the order of operations.

112 Chapter Ten

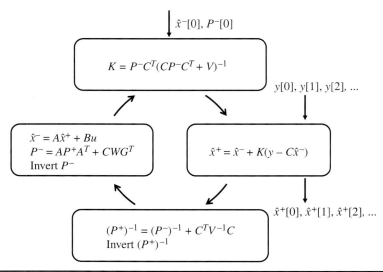

FIGURE 10.6 Alternative Kalman filter loop.

Solved Problems

Problem 10.1
The system is given by

$$x[k+1] = x[k] + w[k]$$
$$y[k] = x[k] + v[k]$$

where $w[k]$ and $v[k]$ are white, independent Gaussian noises. The initial state is normally distributed. Also,

$$E(w[k]) = E(v[k]) = 0$$
$$W = 1, \quad V = 2$$
$$E(x[0]) = \hat{x}[0], \quad E((x[0] - \hat{x}[0])^2) = p[0]$$
$$y[1] = 2, \quad y[2] = 3, \quad y[3] = 1$$

A. Write the Kalman filter equations for this system.
B. Write the equation for P_∞ (error covariance in steady state).
C. Compute the first two estimated states $\hat{x}[1], \hat{x}[2]$.
D. Check the stability of the filter in the steady state for $W = 1$, $V = 0.1$, $p[0] = 1$, $\hat{x}[0] = 0$.

Kalman Filter

Solution

A. Given:

$$\begin{cases} A = 1; B = 0; C = 1; G = 1; W = 1; V = 2 \\ Ex[0] = \hat{x}[0]; E(x[0] - \hat{x}[0])^2 = p[0] \end{cases}$$

Initialization: Let's choose $\hat{x}^-[0] = \hat{x}[0]$. This choice is the best if you don't have any additional information about the initial state. In addition, we choose $P^-[0] = p[0]$. Note that the measurement $y[0]$ is missing; thus, we will not be able to do the first measurement update (we don't get any information about the system's state upfront); thus, we must skip that step and also assume that $\hat{x}^+[0] = \hat{x}[0]$ and $P^+[0] = p[0]$, that is, we start estimating at time $k = 1$.

Kalman filter equations for $k \geq 1$:

Measurement update:

$$K[k] = P^-[k]C^T(CP^-[k]C^T + V)^{-1} = \frac{P^-[k]}{P^-[k] + 2} \tag{10.15}$$

$$\hat{x}^+[k] = \hat{x}^-[k] + K[k](y[k] - C\hat{x}^-[k]) = \hat{x}^-[k] + \frac{P^-[k]}{P^-[k] + 2}(y[k] - \hat{x}^-[k])$$
$$= \frac{P^-[k]y[k] + 2\hat{x}^-[k]}{P^-[k] + 2} \tag{10.16}$$

$$P^+[k] = P^-[k] - K[k]CP^-[k] = P^-[k]\left(1 - \frac{P^-[k]}{P^-[k] + 2}\right) = \frac{2P^-[k]}{P^-[k] + 2} \tag{10.17}$$

Time update:

$$L[k] = AP^-[k]C^T(CP^-[k]C^T + V)^{-1} = K[k] = \frac{P^-[k]}{P^-[k] + 2} \tag{10.18}$$

$$\hat{x}^-[k+1] = A\hat{x}^-[k] + Bu[k] + L[k](y[k] - C\hat{x}^-[k]) = \hat{x}^+[k] \tag{10.19}$$

$$P^-[k+1] = (I - K[k]C)P^-[k]A^T + GWG^T = P^+[k] + 1 \tag{10.20}$$

B. In steady state,

$$P^+[k-1] = P^+[k](= P_\infty) \tag{10.21}$$

and we could define the error covariance as $P_\infty = \lim_{k \to \infty} P^+[k]$.

From (10.17) and (10.20) we get: $P_\infty = \frac{2(P_\infty + 1)}{P_\infty + 3} \Rightarrow P_\infty = 1$; thus, the solution is positive definite.

C. From (10.19) and (10.20) we get: $P^-[1] = P^+[0] + 1 = p[0] + 1$; $\hat{x}^-[1] = \hat{x}[0]$, thus for $k = 1$:

$$K[1] = \frac{p[0]+1}{p[0]+3}$$

$$\hat{x}^+[1] = \frac{(p[0]+1)2 + 2\hat{x}[0]}{p[0]+3} = \hat{x}[0] + K[1](y[1] - \hat{x}[0])$$

$$P^+[1] = \frac{2(p[0]+1)}{p[0]+3}$$

$$L[1] = K[1] = \frac{p[0]+1}{p[0]+3}$$

$$\hat{x}^-[2] = \hat{x}^+[1] = \frac{(p[0]+1)2 + 2\hat{x}[0]}{p[0]+3}$$

$$P^-[2] = P^+[1] + 1 = \frac{2(p[0]+1)}{p[0]+3} + 1 = \frac{3p[0]+5}{p[0]+3}$$

Similarly, for $k = 2$:

$$K[2] = \frac{\frac{3p[0]+5}{p[0]+3}+1}{\frac{3p[0]+5}{p[0]+3}+3} = \frac{2p[0]+4}{3p[0]+7}$$

$$\hat{x}^+[2] = \frac{(p[0]+1)2 + 2\hat{x}[0]}{p[0]+3} + \frac{2p[0]+4}{3p[0]+7}\left(y[2] - \frac{(p[0]+1)2+2\hat{x}[0]}{p[0]+3}\right)$$

$$P^+[2] = \frac{3p[0]+5}{p[0]+3}\left(1 - \frac{2p[0]+4}{3p[0]+7}\right)$$

$$L[2] = K[2] = \frac{2p[0]+4}{3p[0]+7}$$

$$\hat{x}^-[3] = \hat{x}^+[2]$$

$$P^-[3] = P^+[2] + 1$$

The same way you continue for $k > 2$.

D. From (10.16) and (10.19):

$$\begin{cases} \hat{x}^+[k] = \hat{x}^-[k] + K[k](y[k] - \hat{x}^-[k]) \\ \hat{x}^-[k] = \hat{x}^+[k-1] \end{cases}$$

Thus, $\hat{x}^+[k] = \hat{x}^+[k-1] + K[k](y[k] - \hat{x}^+[k-1]) = (1 - K[k])\hat{x}^+[k-1] + K[k]y[k]$. For this system to be stable, we need the eigenvalue of $1 - K_\infty$ to be in the unit circle.

Similar to part (B), we can compute the covariance matrix of the error for new data (notice that now $V = 0.1$). We get $P_\infty = \dfrac{0.1(P_\infty + 1)}{P_\infty + 1.1} \Rightarrow P_\infty = 0.0916$, thus $K_\infty \underset{(10.15)+(10.20)}{=} \dfrac{P_\infty + 1}{P_\infty + 1.1} = 0.916$. Finally, $\det(z - 1 + K_\infty) = z - 0.084$, which means that the system is stable.

Problem 10.2

In this problem, we will use simple physical experiment shown in Figure 10.1 to demonstrate optimal estimation. The goal of the experiment is to estimate the value of gravitational constant g from the measurements obtained for the falling ball. The ball location in time is given by $y(t) = \dfrac{gt^2}{2}$. The measurements are given by $(t[k], y[k])$, $1 \leq k \leq K$, where $t[k]$ is the time of measurement and $y[k]$ is the distance from the initial point measured by a sensor. So, $y(t = 0) = 0$, and after that there is a free fall. The sensor is not perfect, so it has noisy measurements. The noise is modeled as zero mean, Gaussian, and with standard deviation σ^2. Find the approximation of g using the first few steps in the Kalman recursion.

Solution

As you know that in free fall after time t, the ball is passing the distance $y = \dfrac{gt^2}{2}$. Let's assume that we receive noisy measurements of the ball location $y[k]$ at times $t[k]$. We will define the state variable we would like to estimate: $x = g = $ const, and assume (for simplicity) that there is no system noise. We will get the following state-space system:

$$\begin{cases} x[k+1] = Ax + Bu + Gw = x[k] \\ y[k] = Cx + v = \underbrace{\left(\dfrac{1}{2} t^2[k]\right)}_{C[k]} x[k] + v[k] \end{cases}$$

From the system above and other data given in the problem, we identify the following variables:

$$A[k] = 1;\ B[k] = 0;\ C[k] = \dfrac{1}{2} t^2[k];\ G[k] = 0;\ W[k] = 0;\ V[k] = \sigma^2$$

Initialization: Let's choose $\hat{x}^-[0] = 0$ (or you can choose $\hat{x}^-[0] = 10$ to get it closer to physical reality and for faster convergence). We choose $P^-[0] = p[0]$ (we don't have any information on estimation error; thus, $p[0]$ should be large).

Measurement update:

$$K[0] = P^-[0] C^T[0] (C[0] P^-[0] C[0]^T + V)^{-1} = \dfrac{P^-[0] C[0]^T}{C[0] P^-[0] C[0]^T + V}$$

$$= \dfrac{1}{(C[0] + V/(C[0] P^-[0]))} = \dfrac{2t^2[0] p[0]}{t^4[0] p[0] + 4V} \underset{t[0]=0}{=} 0$$

$$\hat{x}^+[0] = \hat{x}^-[0] + K[0](y[0] - C[0]\hat{x}^-[0]) = 0 + 0\left(y[0] - \dfrac{t[0]^2}{2} \cdot 0\right) = 0$$

$$P^+[0] = P^-[0] - K[0] C[0] P^-[0] = p[0] - 0 = p[0]$$

Time update:

$$\hat{x}^-[1] = A\hat{x}^+[0] + Bu[0] = \hat{x}^+[0] = 0$$

$$P^-[1] = A(I - K[0]C[0])P^-[0]A^T + GWG^T = (1-0)p[0] = p[0]$$

Similarly, for $k = 1$:
Measurement update:

$$K[1] = \frac{1}{(C[1] + V/(C[1]P^-[1]))} = \frac{2t[1]^2 p[0]}{t[1]^4 p[0] + 4V}$$

$$\hat{x}^+[1] = K[1](y[1] - 0) = \frac{2t[1]^2 p[0] y[1]}{t[1]^4 p[0] + 4V}$$

$$P^+[1] = p[0] - K[1]\frac{t[1]^2}{2}p[0] = p[0] - 0 = p[0]\left(1 - \frac{t[1]^4 p[0]}{t[1]^4 p[0] + 4V}\right)$$

Time update:

$$\hat{x}^-[2] = A\hat{x}^+[1] + Bu[1] = \hat{x}^+[1] = \frac{2t[1]^2 p[0] y[1]}{t[1]^4 p[0] + 4V}$$

$$P^-[2] = A(I - K[1]C[1])P^-[1]A^T + GWG^T = P^+[1] = \left(1 - \frac{t[1]^4 p[0]}{t[1]^4 p[0] + 4V}\right)p[0]$$

We continue for $k > 1$ in the same way.

Question: Check what happens to the estimated state $\hat{x}^+[1]$ when $p[0] \to \infty$ by computing the limit. Was the result expected?

Answer: The computation is simple:

$$x^+[1] = \lim_{p[0] \to \infty} \frac{2t[1]^2 p[0] y[1]}{t[1]^4 p[0] + 4V} = \frac{2y[1]}{t[1]^2}$$

This result is expected because huge uncertainty about the initial ball position makes the model useless. Thus, we have to trust the measurement and twice the measurement $y[1]$ divided by the time squared would show something very close to the acceleration constant g $\left(y = \frac{gt^2}{2} \to g = \frac{2y}{t^2} \text{ and } x = g\right)$.

Question: Now, try to substitute $V \to 0$ in the formulas of $\hat{x}^{\pm}[k]$. Does the result make sense?

Answer: If $V \to 0$, this means that the measurement is perfect and there is no need of alternatives. As in the previous question, the result of estimation will be based on measurement only.

Problem 10.3

The system is given by

$$x[k+1] = 0.5x[k] + w[k]$$

where $x[0]$, $w[k]$ are distributed normally (Gaussian distribution): $Ew[k] = Ex[0] = 0$, $Ew[k]^2 = Ex[0]^2 = 1$.

If there are no measurements,

- A. List the updated equations for $Ex[k]$ and $Cov(x[k])$ by writing explicitly $x[k]$ as a function of $w[j]$ and $x[0]$.
- B. Obtain the same equations from (A) using the Kalman filter equations. (Hint: "no measurement" is equivalent to $V \to \infty$.)

Now, the measurement is given by $y[k] = x[k] + v[k]$, where $v[k]$ is white Gaussian noise: $Ev[k] = 0$, $Ev[k]^2 = 1$.

- C. Write the equations for $Ex[1]$, given $y[0]$, $y[1]$.
- D. Write the Kalman equation for $\hat{x}[k] = E(x[k])|_{k \geq 2}$ given $y[0]$, $y[1]$ if only two measurements $y[0]$, $y[1]$ are available.

Solution

A. Let's explicitly compute a few iterations for the system's states:

$$x[1] = 0.5x[0] + w[0]$$

$$x[2] = 0.5x[1] + w[1] = 0.25x[0] + 0.5w[0] + w[1]$$

$$x[3] = 0.5x[2] + w[2] = 0.125x[0] + 0.25w[0] + 0.5w[1] + w[2]$$

It is easy to pick up a general pattern and write

$$x[k] = \left(\frac{1}{2}\right)^k x[0] + \sum_{i=0}^{k-1} \left(\frac{1}{2}\right)^i w[k-i-1]$$

Then, the expected value is

$$E(x[k]) = E\left(\left(\frac{1}{2}\right)^k x[0] + \sum_{i=0}^{k-1} \left(\frac{1}{2}\right)^i w[k-i-1]\right)$$

$$= \left(\frac{1}{2}\right)^k E(x[0]) + \sum_{i=0}^{k-1} \left(\frac{1}{2}\right)^i E(w[k-i-1]) = 0$$

since $E(x[0]) = E(w[k-i-1]) = 0$.

To compute the covariance, or rather variance due to the scalar nature of the given system, we use the formula $Var(X) = E\{X^2\} - (E\{X\})^2 = E\{X^2\}$:

$$Var(x[k]) = E\left\{\left[\left(\frac{1}{2}\right)^k x[0] + \sum_{i=0}^{k-1}\left(\frac{1}{2}\right)^i w[k-i-1]\right]^2\right\}$$

$$= E\left\{\left(\frac{1}{2}\right)^{2k} x[0]^2 + \left(\sum_{i=0}^{k-1}\left(\frac{1}{2}\right)^i w[k-i-1]\right)^2 + 2\left(\frac{1}{2}\right)^k x[0]\sum_{i=0}^{k-1}\left(\frac{1}{2}\right)^i w[k-i-1]\right\}$$

$$= E\left\{\left(\frac{1}{2}\right)^{2k} x[0]^2 + \left(\sum_{i=0}^{k-1}\left(\frac{1}{2}\right)^i w[k-i-1]\right)^2\right\}$$

$$= \left(\frac{1}{2}\right)^{2k} Var(x[0]) + \sum_{i=0}^{k-1}\left(\frac{1}{2}\right)^{2i} Var(w[k-i-1])$$

$$= \left(\frac{1}{4}\right)^k + \sum_{i=0}^{k-1}\left(\frac{1}{4}\right)^i = \sum_{i=0}^{k}\left(\frac{1}{4}\right)^i = \frac{1-\left(\frac{1}{4}\right)^{k+1}}{1-\frac{1}{4}} = \frac{4}{3} - \frac{1}{3}\left(\frac{1}{4}\right)^k$$

B. The Kalman equations for $V \to \infty$:

$$K[k] = P^-[k]C^T(CP^-[k]C^T + V)^{-1} = 0$$

$$\hat{x}^+[k] = \hat{x}^-[k] + K[k](y[k] - C\hat{x}^-[k]) = \hat{x}^-[k]$$

$$P^+[k] = P^-[k] - K[k]CP^-[k] = P^-[k]$$

$$L[k] = AK[k] = 0$$

$$\hat{x}^-[k+1] = A\hat{x}^+[k] + Bu[k] = A\hat{x}^-[k] + Bu[k] + L[k](y[k] - C\hat{x}^-[k]) = 0.5\hat{x}^+[k]$$

$$P^-[k+1] = A(I - K[k]C)P^-[k]A^T + GWG^T = 0.5P^-[k]0.5 + 1$$

$$P[0] = 1$$

Thus, $\hat{x}^+[k+1] = (0.5)^k \hat{x}[0]$ and $P^-[k+1] = 0.25P^-[k] + 1$; therefore, $P[k] = 1 + 0.25 \times (1 + 0.25(1 + 0.25(1 + 0.25(...)))) = \sum_{i=0}^{k}\left(\frac{1}{4}\right)^i$ and the results are the same as in (A).

C. The equations are

$$K[0] = \frac{P^-[0]C^T}{CP^-[0]C^T + R} = \frac{1}{1+1} = \frac{1}{2}$$

Thus, $E(x[0])$ given $y[0]$ is $x^-[0] + K[0](y[0] - x^-[0]) = \dfrac{y[0]}{2}$

$$P^+[0] = (1 - K[0]C)P^-[0] = \dfrac{1}{2}$$

$$P^-[1] = AP^+[0]A^T + Q[0] = \left(\dfrac{1}{2}\right)^3 + 1 = \dfrac{9}{8}$$

$$K[1] = P^-[1](P^-[1] + R[1])^{-1} = \dfrac{9}{17}$$

Here, $E(x[1])$ given $y[0]$, $y[1]$ is $\hat{x}^+[1] = x^-[1] + K[1](y[1] - x^-[1]) = \dfrac{y[0]}{4} + \dfrac{9}{17} \times \left(y[1] - \dfrac{y[0]}{4}\right) = \dfrac{2y[0] + 9y[1]}{17}$.

D. The Kalman equation is
$$\hat{x}_k = \dfrac{1}{2}\hat{x}_{k-1}$$

CHAPTER 11
Linear Quadratic Gaussian Control

We have so far discussed optimal controllers and optimal observers. Now it is time to put them together into a single closed-loop system. This should not come as a big surprise for us, since we already know about the separation principle. The system with a linear quadratic regulator (LQR)–based controller and a Kalman filter is called a linear quadratic Gaussian (LQG) control system. The only change that we will see in this chapter is the continuous-time version of the Kalman filter (this is the Kalman-Bucy filter). As you may expect, the solution of the Kalman-Bucy equations is dual to the solution of continuous LQR (with some changes in matrix names).

Kalman-Bucy Filter

The *Kalman-Bucy filter* is a continuous-time version of the Kalman filter and represents an optimal state estimator (observer) discussed in Chapter 10.

The system is given by

$$\begin{cases} \dot{x} = Ax + Bu + Gw \\ \tilde{y} = Cx + Du + v \end{cases}$$

where v and w are white Gaussian noises. In most cases, their covariance matrices V and W for these noises are diagonal. The system is controlled by the input u, and w models the disturbances in the system.

Theorem 11.2 shows the relation to optimal regulator LQR.

Theorem 11.2 (Duality Principle)

The solutions of LQR and the Kalman-Bucy filter are dual. One needs to do the following substitutions in the Riccati and other equations to find the optimal state observer:

LQR	A	B	S	Q	R	K
Kalman	A^T	C^T	P	GWG^T	V	L^T

In other words, following Equations (8.2) and (8.3), the optimal observer is given by

$$L = PC^T V^{-1} \tag{11.1}$$

where P is the solution of the Riccati equation:

$$AP + PA^T + GWG^T - PC^T V^{-1} CP = 0 \tag{11.2}$$

What Is LQG Control?

LQG control is a modern method for finding an optimal dynamic regulator. It allows providing necessary performance, while keeping the control effort low. With this method, it is possible to improve the robustness to noise. The problem of regulation regarding the measurement noise v and the system's noise w is described in Figure 11.1. Note that in this chapter we consider only the so-called *infinite horizon LQG* that has constant controller and observer gains. Full description of the LQG approach is beyond the scope of this book.

Our goal is to minimize y to zero. The LQG regulator measures the noisy output $\tilde{y} = y + v$ and creates the control effort u.

Theorem 11.2 (Separation Principle)

To minimize the performance criterion $\tilde{J} = E\left\{\int_0^T (x^T Q x + 2 x^T N u + u^T R u) dt\right\}$ (E denotes expected value or average), we do the following:

1. Design an LQR controller for the deterministic problem (as there is no noise) and compute the matrix of the controller's gains K using Formulas (8.6) and (8.7):

$$A^T S + SA + Q - (SB + N)R^{-1}(SB + N)^T = 0$$

$$K = R^{-1}(B^T S + N^T)$$

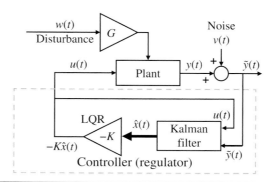

FIGURE 11.1 Block diagram of the LQG regulator.

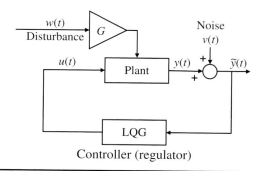

FIGURE 11.2 A general block diagram of the LQG regulator.

2. Design an optimal observer (a Kalman-Bucy filter) for the given system and compute the matrix of gains L using Formulas (11.1) and (11.2).
3. Use the control effort $u = -K\hat{x}$, where $\hat{x}$ is the output of the Kalman-Bucy filter $\dot{\hat{x}} = A\hat{x} + Bu + L(\tilde{y} - C\hat{x} - Du)$, and $E(ww^T) = W; E(vv^T) = V; E(wv^T) = N$.

In other words, we design the LQR assuming that all states are measurable and there is no noise, and then we independently design the Kalman-Bucy filter assuming that the system is stable. Finally, we connect both designs together and get the LQG controller. A general block diagram of the closed-loop system is shown in Figure 11.2.

The state equations of the LQG-controlled system:

$$\begin{cases} \dot{\hat{x}} = [A - LC - (B - LD)K]\hat{x} + L\tilde{y} \\ u = -K\hat{x} \end{cases} \quad (11.3)$$

Since only input and output of the plant are connected to the observer and controller, we could think about them together as a transfer function $F(s)$. From (11.3), this feedback transfer function (*LQG controller transfer function*) is

$$F(s) = -K(sI - A + LC + (B - LD)K)^{-1}L \quad (11.4)$$

NOTE The system is connected with a positive sign in the feedback.

Optimal Cost Function for Stationary LQG

Continuous-time case:

$$J_{min} = tr\{SBKP + SGWG^T\} \quad (11.5)$$

Discrete-time case:

$$J_{min} = tr\{SBKPA^T + SGWG^T\} \quad (11.6)$$

Solved Problems

Problem 11.1
Given the scalar system:

$$\begin{cases} \dot{x} = ax + u + w \\ y = cx + v \end{cases}$$

where a and c are given constants, and the noises w and v are distributed normally with zero mean and variance W and V, respectively.

A. Find the LQG regulator for this system assuming that the optimization criterion is given by $\tilde{J} = E\left\{\int_0^T (qx^2 + \rho u^2) dt\right\}$ with given q and ρ constants.
B. What happens to the system if $\rho \to \infty$?
C. What happens if $W \to 0$?
D. What happens if $W \gg V$?
E. Find the transfer function of the LQG controller for $a = 2$, $c = 1$, $\rho = 1$, $q = 5$, $V = 2$, and $W = 24$.

Solution
A. From the system's definition, $B = G = 1$. Also, $Q = q$ and $R = \rho$. First, we design an LQR controller using Equation (8.3):

$$A^T S + SA + Q - SBR^{-1}B^T S = 0$$

$$aS + Sa + q - \frac{S^2}{\rho} = 0$$

$$S^2 - 2a\rho S - q\rho = 0$$

There are two solutions to this quadratic equation, but we will choose only one positive solution:

$$S = \frac{2a\rho + \sqrt{4a^2\rho^2 + 4q\rho}}{2} = a\rho + \sqrt{a^2\rho^2 + q\rho}$$

From (8.2):

$$K = R^{-1}B^T S$$

$$K = \frac{S}{\rho} = \frac{1}{\rho}(a\rho + \sqrt{a^2\rho^2 + q\rho}) = a + \frac{\sqrt{a^2\rho^2 + q\rho}}{\rho} = a + \sqrt{a^2 + \frac{q}{\rho}}$$

Similarly, we design the Kalman-Bucy stationary filter using (11.2):

$$AP + PA^T + GWG^T - PC^TV^{-1}CP = 0$$

$$aP + Pa + W - \frac{P^2c^2}{V} = 0$$

$$c^2P^2 - 2aVP - VW = 0$$

There are two solutions to this quadratic equation, but we will choose only one positive solution:

$$P = \frac{2aV + \sqrt{4a^2V^2 + 4c^2WV}}{2c^2} = \frac{aV + \sqrt{a^2V^2 + c^2WV}}{c^2}$$

From (11.1):

$$L = PC^TV^{-1} = \left(\frac{aV + \sqrt{a^2V^2 + c^2WV}}{c^2}\right)\frac{c}{V} = \frac{a}{c} + \frac{\sqrt{a^2 + c^2W/V}}{c}$$

$$= \frac{a}{c} + \sqrt{\left(\frac{a}{c}\right)^2 + \frac{W}{V}}$$

Now, connecting the system as shown in Equation (11.3) and Figure 11.2 provides the LQG regulator.

B. If $\rho \to \infty$, then $\frac{q}{\rho} \to 0$ and $K \to a + \sqrt{a^2} = a + |a|$. Note that in this case if the system is stable in an open loop ($a < 0$), then $K = 0$. In other words, the LQR is not changing the position of the open-loop pole and leaves it in the closed loop at a, as expected from the symmetric root locus. If the system is unstable ($a > 0$), we would expect the unstable pole jump to its mirror image location in the left complex semi-plane. Let's see what is going on. If $a > 0$, then $K = 2a$, and the location of the closed-loop pole will be $\det(sI - A + BK) = \det(s - a + 1 \cdot 2a) = \det(s + a) = s + a$ (for scalar quantities). Thus, the optimal closed-loop location is at $-a$ as expected.

C. If $W \to 0$, then $L = \frac{a}{c} + \left|\frac{a}{c}\right|$. This means that if a and c are of a different sign, then $L = 0$, and if they are of the same sign, then $L = \frac{2a}{c}$. For example, if the system is stable ($a < 0$) and $c > 0$, then the best thing to do would be trusting the system and ignoring the output.

D. We can rewrite $L = \frac{a}{c} + \sqrt{\left(\frac{a}{c}\right)^2 + \frac{W}{V}} = \frac{a}{c}\left(1 + \sqrt{1 + \frac{W}{V}\left(\frac{c}{a}\right)^2}\right)$. If $W \gg V$, it follows that $\frac{W}{V}\left(\frac{c}{a}\right)^2 \gg 1$ and $\sqrt{1 + \frac{W}{V}\left(\frac{c}{a}\right)^2} \to \sqrt{\frac{W}{V}\left(\frac{c}{a}\right)^2}$. For the same reason, $\sqrt{\frac{W}{V}\left(\frac{c}{a}\right)^2} \gg 1$, thus $1 + \sqrt{\frac{W}{V}\left(\frac{c}{a}\right)^2} \approx \sqrt{\frac{W}{V}\left(\frac{c}{a}\right)^2}$. In conclusion, $L \approx \frac{a}{c}\sqrt{\frac{W}{V}\left(\frac{c}{a}\right)^2} = \sqrt{\frac{W}{V}}$.

E. Using (11.4):

$$F(s) = -K(sI - A + LC + (B - LD)K)^{-1}L$$

$$F(s) = -\left(a + \sqrt{a^2 + \frac{q}{\rho}}\right)\left(s - a + \left(a + \sqrt{a^2 + \frac{Wc^2}{V}}\right) + a + \sqrt{a^2 + \frac{q}{\rho}}\right)^{-1}\left(\frac{a}{c} + \sqrt{\left(\frac{a}{c}\right)^2 + \frac{W}{V}}\right)$$

$$= -\left(2 + \sqrt{2^2 + \frac{5}{1}}\right)\left(s - 2 + 2 + \sqrt{2^2 + 24 \cdot \frac{1^2}{2}} + 2 + \sqrt{2^2 + \frac{5}{1}}\right)^{-1}\left(\frac{2}{1} + \sqrt{\left(\frac{2}{1}\right)^2 + \frac{24}{2}}\right)$$

$$= -\frac{-5 \cdot 6}{s + 9} = -\frac{30}{s + 9}$$

The controller is a simple first-order low-pass filter.

Problem 11.2
A system is shown in Figure 11.3.
   Design an LQG controller for the following criterion:

$$V = E(v^2) = 0.1; \quad W = E(w^2) = I$$

$$J = \int_0^\infty (4y^2 + 0.1u^2)dt$$

Would the design change if we replace J with $\int_0^\infty (40y^2 + u^2)dt$?

Solution
The plant transfer function is minimal, thus observable and controllable. So, LQG can be designed.
   Let's start with the canonical controller realization of our plant:

$$A = \begin{pmatrix} -2 & -36 \\ 1 & 0 \end{pmatrix}; \quad B = \begin{pmatrix} 1 \\ 0 \end{pmatrix}; \quad C = (0 \quad 36); \quad D = 0$$

NOTE You could choose any other equivalent realization. If you choose another realization, the solution for the Riccati equations, L and K, will be different, but the final result $F(s)$ should be the same since the optimal controller is unique.

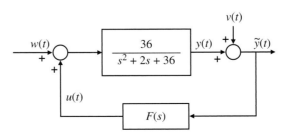

FIGURE 11.3 System for Problem 11.2.

We need to write $J = \int_0^\infty (5y^2 + 0.1u^2)dt$ as $J = \int_0^\infty (x^T Q x + R u^2)dt$. Since $y = Cx$, we get $4y^2 = 4(Cx)^T(Cx) = x^T 4C^T C x$, thus $Q = 4C^T C = 4\begin{pmatrix} 0 \\ 36 \end{pmatrix}(0 \quad 36) = \begin{pmatrix} 0 & 0 \\ 0 & 5184 \end{pmatrix}$. The coefficient of u^2 is $R = 0.1$.

Now, we have all components to solve LQR's Riccati Equation (8.3):

$$A^T S + SA + Q - SBR^{-1}B^T S = 0$$

$$\begin{pmatrix} -2 & 1 \\ -36 & 0 \end{pmatrix}\begin{pmatrix} s_1 & s_2 \\ s_2 & s_3 \end{pmatrix} + \begin{pmatrix} s_1 & s_2 \\ s_2 & s_3 \end{pmatrix}\begin{pmatrix} -2 & -36 \\ 1 & 0 \end{pmatrix} + \begin{pmatrix} 0 & 0 \\ 0 & 5184 \end{pmatrix} - \begin{pmatrix} s_1 & s_2 \\ s_2 & s_3 \end{pmatrix}\begin{pmatrix} 1 \\ 0 \end{pmatrix} 10 (1 \quad 0)\begin{pmatrix} s_1 & s_2 \\ s_2 & s_3 \end{pmatrix}$$

$$= \begin{pmatrix} 0 & 0 \\ 0 & 0 \end{pmatrix}$$

This system is equivalent to the following system of equations:

$$\begin{cases} -2s_1 + s_2 - 2s_1 + s_2 - 10s_1^2 = 0 \\ -2s_2 + s_3 - 36s_1 - 10s_1 s_2 = 0 \\ -36s_2 - 36s_2 + 5184 - 10s_2^2 = 0 \end{cases}$$

From the third quadratic equation $-36s_2 - 36s_2 + 5184 - 10s_2^2 = 0$, it follows that $5s_2^2 + 36s_2 - 2592 = 0$, and the two possible solutions are

$$s_2 = \frac{-36 \pm \sqrt{36^2 - 4 \cdot 5 \cdot 2592}}{2 \cdot 5} = 19.4512 \quad \text{or} \quad -26.6512$$

If we pick the positive solution $s_2 = 19.4512$ and substitute it in the first equation $-2s_1 + s_2 - 2s_1 + s_2 - 10s_1^2 = 0$, we get the quadratic equation

$$10s_1^2 + 4s_1 - 38.902 = 0$$

for which we have two solutions: $s_1 = 1.7825$ and $s_1 = -2.1825$. Using the Sylvester criterion, we have to choose positive s_1 to make S positive definite. Similarly, when substituting the result in the second equation, we get the only solution $s_3 = 449.7877$.

The only positive definite solution of this system of equations is $S = \begin{pmatrix} 1.7825 & 19.4512 \\ 19.4512 & 449.7877 \end{pmatrix}$.

Based on (8.2), the LQR stationary gain will be

$$K = R^{-1}B^T S = 10(1 \quad 0)\begin{pmatrix} 1.7825 & 19.4512 \\ 19.4512 & 449.7877 \end{pmatrix} = (17.825 \quad 194.512)$$

Now, we can design the Kalman-Bucy filter using (11.2):

$$AP + PA^T + GWG^T - PC^T V^{-1} CP = 0$$

$$\begin{pmatrix} -2 & -36 \\ 1 & 0 \end{pmatrix}\begin{pmatrix} p_1 & p_2 \\ p_2 & p_3 \end{pmatrix} + \begin{pmatrix} p_1 & p_2 \\ p_2 & p_3 \end{pmatrix}\begin{pmatrix} -2 & 1 \\ -36 & 0 \end{pmatrix} + \begin{pmatrix} 1 & 0 \\ 0 & 1 \end{pmatrix} - \begin{pmatrix} p_1 & p_2 \\ p_2 & p_3 \end{pmatrix}\begin{pmatrix} 0 \\ 36 \end{pmatrix} 10 (0 \quad 36)\begin{pmatrix} p_1 & p_2 \\ p_2 & p_3 \end{pmatrix}$$

$$= \begin{pmatrix} 0 & 0 \\ 0 & 0 \end{pmatrix}$$

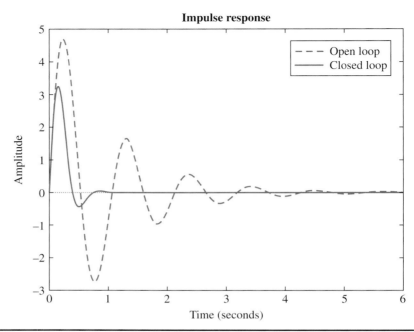

FIGURE 11.4 Open- and closed-loop impulse response of the system in Problem 11.2.

which is equivalent to

$$\begin{cases} -2p_1 - 36p_2 - 2p_1 - 36p_2 + 1 - 12,960p_2^2 = 0 \\ -2p_2 - 36p_3 + p_1 - 12,960p_2p_3 = 0 \\ p_2 + p_2 + 1 - 12,960p_3^2 = 0 \end{cases}$$

You can directly check that the only positive definite solution of this system is

$$P = \begin{pmatrix} 0.2582 & -0.0005 \\ -0.0005 & 0.0088 \end{pmatrix}$$

and the optimal stationary observer is (11.1)

$$L = PC^T V^{-1} = \begin{pmatrix} -0.18 \\ 3.16 \end{pmatrix}$$

The impulse responses in open and closed loops are shown in Figure 11.4.

The design will not change if we multiply cost J by any positive constant, since addition and multiplication by a constant move the cost function up or down without changing the location of its minimum.

CHAPTER 12
Project Examples

In this chapter, we present a few practical control problems. We assume that you will use MATLAB and Simulink to solve the problems and produce simulations of the system, but other computational tools could be used as well.

General Instructions

1. Before you start working on a project, you should brush up your knowledge of different MATLAB Control Toolbox commands that may help you.
2. All models, codes, computations, graphs, and simulations, together with the appropriate explanation, must be included while submitting the work. The goal is that your results would be reproducible based on your report alone.
3. Pay attention to careful titling your graphs and graph axes. You could add any annotations to your graphs (as needed). Specify and emphasize all important points. If there is more than one curve on the graph, you have to name each one and explain the trend.
4. Write all your conclusions explicitly; do not leave space for guessing.
5. All final answers must be emphasized.
6. It is necessary to maintain an orderly work.

Project 1: Magnetic Levitation System Control

Levitation System Model
The goal of the levitation system is to hold a small metallic ball in the air using magnetic force (Figure 12.1).

The system should control the electromagnetic field to keep the ball at some pre-defined distance from the electromagnet. In an ideal case, we just need to increase the electromagnetic force until it is equal to the gravitational force, but in real world there are many disturbances and noise, so it is impossible to create a perfect electromagnetic field. These disturbances will cause the ball to fall or stick to the electromagnet. Thus, we need to develop a closed-loop system to control the ball's height.

System's Description
The system consists of an electromagnet, a metallic ball, a light source, and a detector (see Figure 12.1).

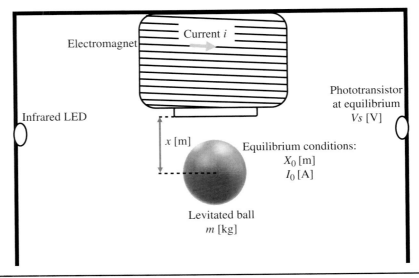

FIGURE 12.1 Magnetic levitation system.

The equations describing the ball's motion in the electromagnetic field are nonlinear. The important variables are

1. The distance between the ball and the magnet: x
2. The current that passes through the electromagnet coil: i
3. The electromagnetic force: f

Motion Equations
Force is given by

$$f = -\frac{i^2}{2}\frac{dL}{dx} \tag{12.1}$$

where L is total inductance of the system. The floating ball is also contributing to the inductivity of the system. When the ball rises, the total inductance L rises too. When the ball falls, the total inductivity decreases, and approaches its minimal value when the ball is far away from the magnet. The minimal value equals to the inductance of the electromagnet coil itself:

$$L = L_1 + \frac{L_0 X_0}{x} \tag{12.2}$$

where L_1 is the coil inductance and L_0 is the inductance that adds the ball at the equilibrium point X_0. Using Equations (12.1) and (12.2), we get

$$f = -\frac{i^2}{2}\left(\frac{-L_0 X_0}{x^2}\right) = \frac{L_0 X_0}{2}\left(\frac{i}{x}\right)^2 \tag{12.3}$$

and at the equilibrium point:

$$f_0 = mg = \frac{L_0 X_0}{2}\left(\frac{I_0}{X_0}\right)^2 \quad (12.4)$$

from which the mass of the ball m could be computed.

The equation of dynamic motion could be written using the second law of Newton, connecting the acceleration and mass to the sum of forces $m\ddot{x} = -(f - f_0)$, where m is the mass of the ball.

If we rearrange the last three equations, we get

$$\ddot{x} = \frac{L_0 X_0}{2m}\left(\left(\frac{I_0}{X_0}\right)^2 - \left(\frac{i}{x}\right)^2\right) = g - \frac{L_0 X_0}{2m}\left(\frac{i}{x}\right)^2 \quad (12.5)$$

Electrical Model

Assume that the electromagnet can be modeled by the resistor and total inductance in series as in Figure 12.2.

Using Kirchhoff's law,

$$v = Ri + L\frac{di}{dt} \quad (12.6)$$

$$\frac{di}{dt} = \frac{v - Ri}{L} = \frac{v - Ri}{L_1 + \frac{L_0 X_0}{x}} \quad (12.7)$$

and after some simple algebraic manipulations on the two equations above we have

$$\dot{i} = \frac{di}{dt} = \frac{x}{L_1 x + L_0 X_0}(v - Ri) \quad (12.8)$$

Sensor Model

The feedback sensor consists of a light source and a light detector. The detector is connected in a way that when the ball falls, the voltage at the sensor's output rises. We assume the simplest gain model for that kind of sensor is

$$v_s = \beta x \quad (12.9)$$

where v_s is the output voltage and β is the gain.

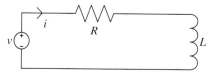

FIGURE 12.2 Electrical model of an electromagnet.

Physical Model

Assume that $L_1 \gg L_0$, and the following values describe our physical model:

Parameter	Value
Equilibrium distance X_0	0.05 [m]
Equilibrium current I_0	7 [A]
Coil resistance R	2.5 [Ω]
Coil inductance L_1	0.2 [Hy]
Ball inductance L_0	0.005 [Hy]
Sensor's gain β	400 [V/m]
Standard gravity g	9.8 [N/kg]

Questions

1. Linearize Equations (12.5) and (12.8) with regard to the equilibrium point $\{X_0, I_0\}$. Define the following state variables: $x_1 = x, x_2 = i, x_3 = \dot{x}$. The input is $u = v$ and the output is $y = v_s$.

   Write the state equations (include the computation). Linearize the obtained system and prove that the linear system $\{A, B, C, D\}$ is described by

   $$A \underset{L_1 \gg L_0}{\simeq} \begin{pmatrix} 0 & 0 & 1 \\ 0 & \frac{-R}{L_1} & 0 \\ \frac{2g}{X_0} & \frac{-2g}{I_0} & 0 \end{pmatrix}; \quad B \underset{L_1 \gg L_0}{\simeq} \begin{pmatrix} 0 \\ \frac{1}{L_1} \\ 0 \end{pmatrix}; \quad C \underset{L_1 \gg L_0}{\simeq} (\beta \ \ 0 \ \ 0); \quad D = 0$$

2. Find the system's transfer function. (From now on, you are free to use MATLAB in any computations, but you have to submit all the code and the produced results.)

3. Check the controllability, observability, stabilizability, and detectability of the system.

4. Implement and show the linear and nonlinear systems in Simulink.

5. Assume that you have access to all the state variables. Design state-space controller for the linear system. Connect that controller to the linear system and test it with at least two initial conditions to show that your design gives stable results (with zero input).

6. Attach the same controller to the nonlinear system (remember to shift all the states and the input to the origin equilibrium point). Show that for some initial conditions the system is stable. Find the initial conditions for which the system is unstable.

7. Now, assume that you don't have access to state variables (only to the input and output). Design the appropriate observer (with zero initial conditions). Connect the observer to a linear system with a controller and check how the convergence is dependent on the initial conditions. In addition to state graphs, show the plant input graph. Is the control effort reasonable with your chosen location of poles?

8. Design LQR controller for the linear system. Choose matrix Q to be diagonal. Play with the values in Q and R to see how control effort and state convergence to zero are dependent on those values. Show graphical results to prove your conclusions.

9. Connect one chosen LQR controller to linear and nonlinear systems. Compare the state, output, and input signals (plot the state output in the same coordinate system for linear and nonlinear systems using "hold on" MATLAB function).

10. Connect small step reference signal. How good is your tracking of steady-state error? Correct the problem by adding a constant gain at the input. Now, try sine wave input. Does the output follow the input well?

Project 2: Double Inverted Pendulum

System Description

In this project, we discuss the stabilization of the double inverted pendulum, as it represents a problem that is equivalent to stabilizing human standing. Human body biomechanics could simply be described as a double inverted pendulum while assuming that the neck and knees are locked. The pendulum consists of two rods as described in Figure 12.3.

The lower rod represents the pair of legs up to the pelvis, and the upper rod represents the upper torso (head + hands + chest + abdomen), which are modeled here as a rigid body. System's control is done by activating τ_1, τ_2 in the appropriate joint points.

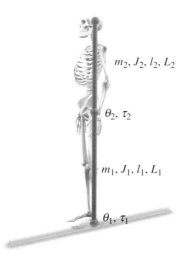

FIGURE 12.3 Double inverted pendulum model of the standing biomechanics.

The system's parameters are as follows:

Parameter	Description	Unit
m_1	Leg mass	kg
m_2	Upper body mass	kg
L_1	Leg length	m
L_2	Upper body length	m
l_1	Distance to the leg mass center	m
l_2	Distance to upper body mass center	m
J_1	Moment of inertia of the legs	kg·m²
J_2	Moment of inertia of the upper body	kg·m²
θ_1	Angle between the legs and the perpendicular from the ground	rad
θ_2	Angle between upper body and legs	rad
θ_3	Sum of angles of the legs and upper body	rad
g	Gravity constant	m/s²
τ_1	Torque powered by ankle	N/m
τ_2	Torque powered by the waist	N/m

Nonlinear Model

To find the nonlinear model, we will use the Euler-Lagrange equation (from calculus of variations and used frequently in analytical mechanics):

$$\frac{\partial}{\partial t}\left[\frac{\partial L}{\partial \dot{q}}\right] - \frac{\partial L}{\partial q} = Q_q \tag{12.10}$$

where $L = T - V$
  T = kinetic energy
  V = potential energy
  Q_q = applied momentums
  Q = generalized coordinates

The overall kinetic energy of the system is given by the sum of the kinetic energies of each of the rods:

$$T = T_1 + T_2 \tag{12.11}$$

$$T_1 = \frac{1}{2}m_1[(l_1\dot{\theta}_1\cos\theta_1)^2 + (l_1\dot{\theta}_1\sin\theta_1)^2] + \frac{1}{2}J_1\dot{\theta}_1^2 \tag{12.12}$$

$$T_2 = \frac{1}{2}m_2[(L_1\dot{\theta}_1\cos\theta_1 + l_2\dot{\theta}_3\cos\theta_3)^2 + (L_1\dot{\theta}_1\sin\theta_1 + l_2\dot{\theta}_3\sin\theta_3)^2] + \frac{1}{2}J_2\dot{\theta}_3^2 \tag{12.13}$$

The total potential energy is given by the sum of the potential energies of each of the rods:

$$U = U_1 + U_2 \tag{12.14}$$

$$U = m_1gl_1\cos\theta_1 + m_2g[L_1\cos\theta_1 + l_2\cos\theta_3] \tag{12.15}$$

Determining matrices Q_i and q_i is done as follows:

$$q_i = \begin{pmatrix} \theta_1 \\ \theta_3 \end{pmatrix} \text{ and } Q_i = \begin{pmatrix} \tau_1 \\ \tau_2 \end{pmatrix} \qquad (12.16)$$

By substitution into Equation (12.10) for $q_1 = \theta_1$:

$$-g\sin\theta_1(m_1l_1 + m_2L_1) + (m_1l_1^2 + m_2L_1^2 + J_1)\ddot{\theta}_1 + \cdots$$
$$\cdots + m_2l_2L_1[\ddot{\theta}_3\cos(\theta_1 - \theta_3) + \dot{\theta}_3^2\sin(\theta_1 - \theta_3) - \dot{\theta}_1\dot{\theta}_3\sin(\theta_1 - \theta_3)] = \tau_1 \qquad (12.17)$$

And for $q_2 = \theta_3$:

$$-g\sin\theta_3 m_2l_2 + (m_2l_2^2 + J_2)\ddot{\theta}_3 + \cdots$$
$$\cdots + m_2l_2L_1[\ddot{\theta}_1\cos(\theta_1 - \theta_3) - \dot{\theta}_1^2\sin(\theta_1 - \theta_3) + \dot{\theta}_1\dot{\theta}_3\sin(\theta_1 - \theta_3)] = \tau_2 \qquad (12.18)$$

The two equations above could be written as follows:

$$h_1\ddot{\theta}_1 + h_2\ddot{\theta}_3\cos(\theta_1 - \theta_3) + h_2\dot{\theta}_3^2\sin(\theta_1 - \theta_3) - h_3\sin\theta_1 - h_2\dot{\theta}_1\dot{\theta}_3\sin(\theta_1 - \theta_3) = \tau_1 \qquad (12.19)$$

$$h_4\ddot{\theta}_3 + h_2\ddot{\theta}_1\cos(\theta_1 - \theta_3) - h_2\dot{\theta}_1^2\sin(\theta_1 - \theta_3) - h_5\sin\theta_3 + h_2\dot{\theta}_1\dot{\theta}_3\sin(\theta_1 - \theta_3) = \tau_2 \qquad (12.20)$$

where h_i constants are described in the following table:

Constant	Value
h_1	$m_1l_1^2 + m_2L_1^2 + J_1$
h_2	$m_2l_2L_1$
h_3	$m_1l_1g + m_2L_1g$
h_4	$m_2l_2^2 + J_2$
h_5	m_2l_2g

Substituting $\theta_2 = \theta_3 - \theta_1$ in Equations (12.19) and (12.20) is providing two differential equations that describe the nonlinear system's model:

$$h_1\ddot{\theta}_1 + h_2(\ddot{\theta}_1 + \ddot{\theta}_2)\cos\theta_2 - h_2(\dot{\theta}_1 + \dot{\theta}_2)^2\sin\theta_2 + h_2\dot{\theta}_1(\dot{\theta}_1 + \dot{\theta}_2)\sin\theta_2 - h_3\sin\theta_1 = \tau_1 \qquad (12.21)$$

$$h_4(\ddot{\theta}_1 + \ddot{\theta}_2) + h_2\ddot{\theta}_1\cos\theta_2 - h_2\dot{\theta}_1^2\sin\theta_2 - h_2\dot{\theta}_1(\dot{\theta}_1 + \dot{\theta}_2)\sin\theta_2 - h_5\sin(\theta_1 + \theta_2)_1 = \tau_2 \qquad (12.22)$$

After a few algebraic manipulations, we get

$$h_1\ddot{\theta}_1 + h_2(\ddot{\theta}_1 + \ddot{\theta}_2)\cos\theta_2 - h_2\dot{\theta}_2(\dot{\theta}_1 + \dot{\theta}_2)^2\sin\theta_2 - h_3\sin\theta_1 = \tau_1 \qquad (12.23)$$

$$h_4(\ddot{\theta}_1 + \ddot{\theta}_2) + h_2\ddot{\theta}_1\cos\theta_2 - h_2\dot{\theta}_1\dot{\theta}_2\sin\theta_2 - h_5\sin(\theta_1 + \theta_2)_1 = \tau_2 \qquad (12.24)$$

Question 1
Find the system's equilibrium points. Which ones are stable? (No computations are necessary—physical argument is sufficient.)

Linearization
To linearize the nonlinear model around 0 equilibrium point, we use the following approximations (correct only for small angles):

$$\cos\theta = 1 \qquad \sin\theta = \theta \qquad \dot{\theta}^2 \approx 0$$

The following linear equations are obtained:

$$h_1\ddot{\theta}_1 + h_2\ddot{\theta}_1 - h_3\theta_1 + h_2\ddot{\theta}_2 = \tau_1 \tag{12.25}$$

$$h_2\ddot{\theta}_1 + h_4\ddot{\theta}_1 - h_5\theta_1 + h_4\ddot{\theta}_2 - h_5\theta_2 = \tau_2 \tag{12.26}$$

that could be written in a matrix form:

$$\underbrace{\begin{pmatrix} h_1+h_2 & h_2 \\ h_2+h_4 & h_4 \end{pmatrix}}_{H} \begin{pmatrix} \ddot{\theta}_1 \\ \ddot{\theta}_2 \end{pmatrix} - \begin{pmatrix} h_3 & 0 \\ h_5 & h_5 \end{pmatrix} \begin{pmatrix} \theta_1 \\ \theta_2 \end{pmatrix} = \begin{pmatrix} \tau_1 \\ \tau_2 \end{pmatrix} \tag{12.27}$$

Question 2
Define the state vector x, the measurement (output) y, and the input u as follows:

$$x = (\theta_1 \ \dot{\theta}_1 \ \theta_2 \ \dot{\theta}_2)^T, \qquad y = (\theta_1 \ \theta_2)^T, \qquad u = (\tau_1 \ \tau_2)^T$$

Use the system of Equation 12.27 and show that you receive the following linear state-space system $\{A, B, C, D\}$:

$$A = \begin{pmatrix} 0 & 1 & 0 & 0 \\ w_1h_3 + w_2h_5 & 0 & w_2h_5 & 0 \\ 0 & 0 & 0 & 1 \\ w_3h_3 + w_4h_5 & 0 & w_4h_5 & 0 \end{pmatrix}$$

$$B = \begin{pmatrix} 0 & 0 \\ w_1 & w_2 \\ 0 & 0 \\ w_3 & w_4 \end{pmatrix}, \qquad C = \begin{pmatrix} 1 & 0 & 0 & 0 \\ 0 & 0 & 1 & 0 \end{pmatrix}, \qquad D = \begin{pmatrix} 0 & 0 \\ 0 & 0 \end{pmatrix}$$

where $w_1 = \dfrac{h_4}{\det(H)}$

$w_2 = -\dfrac{h_2}{\det(H)}$

$w_3 = -\dfrac{h_2 + h_4}{\det(H)}$

$w_4 = \dfrac{h_1 + h_2}{\det(H)}$

$\det(H) = (h_1 + h_2)h_4 - (h_2 + h_4)h_2$

Finding the Controlled System

For a typical system, the parameter values that describe the system are determined in the following manner, depending on the person's height H (in meters) and the weight of M (in kilograms):

Parameter	Value	Units
M_1	$0.322 \cdot M$	kg
M_2	$0.68 \cdot M$	kg
L_1	$0.53 \cdot H$	m
L_2	$0.47 \cdot H$	m
l_1	$0.58 \cdot L_1$	m
l_2	$0.58 \cdot L_2$	m
J_1	$0.017 \cdot M \cdot H^2$	kg·m²
J_2	$0.14 \cdot M \cdot H^2$	kg·m²
g	9.8	N/kg

For parameters that will be used in this project, you must choose your own height H [m] and weight M [kg] values.

Question 3
Consider matrices $\{A, B, C, D\}$ corresponding to the linear system, after the parameters are set.

1. Is the system controllable? Is it observable?
2. Is the system stable?

From now on, suppose that the system can be controlled by the input $u = \tau_1$ only ($\tau_2 = 0$ is forced).

3. Find the appropriate matrices $\{A, B, C, D\}$.
4. Find transfer functions $\dfrac{\theta_1(s)}{u(s)}$, $\dfrac{\theta_2(s)}{u(s)}$ of the system.
5. Is the system still controllable?
6. Is there any input u such that the system stabilizes on the values $\theta_1 = \theta_2 = 0.1$? How does your answer align with the system's controllability?

Question 4
Implement and present the linear and nonlinear model of the system in an open loop in Simulink and present the response to the initial conditions $\begin{pmatrix} x_1 \\ x_2 \\ x_3 \\ x_4 \end{pmatrix}_0 = \begin{pmatrix} \theta_1 \\ \dot{\theta}_1 \\ \theta_2 \\ \dot{\theta}_2 \end{pmatrix}_0 = \begin{pmatrix} 0.1 \\ 0 \\ 0.1 \\ 0 \end{pmatrix}$ for both systems.

Optimal Control: LQR Design

Question 5
It is required to design a controller that minimizes the following cost function:

$$J = \int_0^\infty [x(t)^T \cdot Q \cdot x(t) + \rho u^2(t)] \cdot dt \qquad (12.28)$$

Given the access to all state variables, compute the controller's gain vector K for the following values of ρ: $[10^{-5}, 10^{-8}, 10^{-11}]$ and for the following matrices $[Q_1, Q_2]$:

$$Q_1 = \begin{pmatrix} 0 & 0 & 0 & 0 \\ 0 & 0 & 0 & 0 \\ 0 & 0 & 1 & 0 \\ 0 & 0 & 0 & 0 \end{pmatrix}, \quad Q_2 = \begin{pmatrix} 1 & 0 & 0 & 0 \\ 0 & 0 & 0 & 0 \\ 0 & 0 & 0 & 0 \\ 0 & 0 & 0 & 0 \end{pmatrix}$$

(6 combinations).

1. How does selecting each matrix $[Q_1, Q_2]$ affect the cost function J?
2. Find closed-loop eigenvalues for each combination of ρ and Q. For each matrix Q, draw the location of closed-loop poles as a function of ρ (symmetric root locus).
3. Implement the optimal linear closed-loop system in Simulink and show the response to the initial conditions in Question 4 for each of the six controllers. In a separate plot, show the control effort u for each one of the state-space controllers.
4. Explain the response dependency to the value of each of the matrices and ρ.
5. Is it possible to make the system converge as fast as we wish by increasing or decreasing the value of ρ? Why? (Hint: How does transfer function affect the change in eigenvalues? Bigger hint: Check the SRL separately for each row in C matrix.)
6. Find the nonlinear system response to the same initial conditions with the same gains K. For which gains is the system able to stabilize? Explain. In cases where the system does not stabilize, find the boundary initial conditions of a kind $(a, 0, a, 0)^T$ for which the nonlinear system is stabilizing.

Observer Design
Because of different constraints, the angular velocity values $[\dot{\theta}_1, \dot{\theta}_2]$ cannot be directly measured, and it is necessary to reconstruct them from the measured $[\theta_1, \theta_2]$ and from the input.

Question 6
1. Find the matrix of the observer's gains L_1 which enables the use of all controllers K calculated in Question 5. What are the eigenvalues of the observer? Note that since there are two outputs, matrix L_1 has dimensions of 4×2 and it could be computed using the MATLAB command "place."
2. Add the observer to the closed-loop system from Question 5 and compare the response to initial conditions from Question 4 for all controllers K

that were computed in Question 5. Set to zero the initial conditions of an observer.

3. We define the observer's error by $e = x - \hat{x}$, where e is a 4×1 vector. For $\rho = 10^{-5}$ and Q_1, plot all four components of the observer's error. Explain the behavior of the error as a function of the observer's eigenvalue's location.

Noise Filtering

Now, we assume that the system has uncertainty, and there are disturbances and noises affecting its response. The model is described in the following way:

$$\dot{x} = Ax + Bu + Gw$$

$$y = Cx + v$$

where G is the identity matrix, C matrix has dimensions 2×4, and w, v are white Gaussian independently distributed noises with zero mean:

$$w \sim N(0_{4 \times 1}, w_{var} \cdot I_{4 \times 4})$$

$$v \sim N(0_{2 \times 1}, v_{var} \cdot I_{2 \times 2})$$

where w_{var} and v_{var} are scalars.

Question 7

In the first part of the question, we set $v_{var} = 10^{-7}$. For w_{var} taking the following values: $[1, 10^{-3}]$, answer the following questions:

1. Are the necessary conditions for the Kalman filter being satisfied?
2. Find the filter L_{Kalman} for each value of w_{var}. Specify the Kalman filter's eigenvalues for each case. Note that L_{Kalman} has dimensions 4×2 and you can use MATLAB command "are" to compute it.
   Build the closed-loop model with noise state controller and Kalman filter in Simulink. Pick one of the controllers from Question 5.
3. For each value w_{var}, find and show the system's response to the initial conditions [0.1, 0, 0.1, 0] for both observers $[L_{Kalman}, L_1]$ (total four combinations).
4. Is the system always stable in the four cases above? When it is stable, which of the two observers gives a better response?
5. The observer's cost function is defined as a quadratic error $J(t) = (x(t) - \hat{x}(t))^T (x(t) - \hat{x}(t))$. Compare the quadratic error of the Kalman filter with L_1 observer for one value of w_{var} for which the response is stable. Explain the differences between the errors obtained by each of the observers.
   Now, set $w_{var} = 10^{-7}$.
6. For v_{var} taking the following values: $[10^{-1}, 10^{-3}]$, repeat Sections 2 to 5 of this question.
7. When is it beneficial to use the Kalman filter? Explain.

Project 3: Bridge Crane Control

Choosing the Model
The selected model is a loading bridge system. The purpose of the bridge is, for example, the loading of containers into the ship. At the beginning, the load mass is the mass of the hook only. The hook must appear above the container in a resting state. Sometimes, the workers at the seaport help curb the hook fluctuations and attach it to the container. In this project, it is assumed that the vibration dumping and hook connection are done automatically. After lifting the container, it moves a certain distance and is transferred to the ship's deck. This movement is controlled by the engine control limited by the power the engine can provide and the safety restrictions. In order to control the system, it is required to add feedback.

Model Description
The system is described in Figure 12.4. The m load mass can vary considerably between the weight of the hook itself and the weight of the hook with the maximum mass that the bridge is capable of transferring. However, the mass of the cart and the length of the cable are not altered.

We use the following additional simplifying assumptions:

- The cable has no mass, it is not flexible, and has a fixed length L.
- There is no friction or gliding between the cart and the bridge.
- You can neglect the engine's dynamics and nonlinearity, which creates a horizontal input force u. (This assumption makes sense only if the controller provides $|u|$ and $|\dot{u}|$ that aren't too large.)

Definitions:

M = the cart mass extended by moment of inertia (transmitted through gears) of the engine

m = load mass

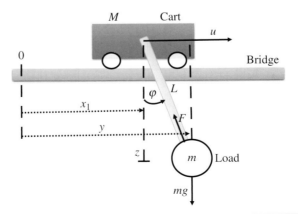

FIGURE 12.4 Bridge crane system.

L = length of cable

g = constant gravitation

x_1 = location of the cart

φ = angle of the cable

y = position of a load

Dynamic Model Equations

Any physical system can be modeled mathematically in one of the following two ways:

A. From measurements of input and output signals. In this case, the system is treated as a "black box" whose internal structure is unknown.

B. From the dynamic relationship between the different parts of the system and their mutual influence. In this case, the structure of the system is known, and the physics laws can be used to construct the model.

We will use the second way to develop nonlinear differential equations that describe the crane bridge system.

First, we examine the system in Figure 12.4 and write down the appropriate force equations. We denote the cable pulling force by F and the vertical distance between the cart and the load by z.

From the equality of forces, the following differential equations could be written:

A. Horizontal forces on a cart:

$$M\ddot{x}_1 = u + F\sin(\varphi) \tag{12.29}$$

B. Horizontal forces on a load:

$$m\ddot{y} = -F\sin(\varphi) \tag{12.30}$$

C. Vertical forces on a load:

$$m\ddot{z} = -F\cos(\varphi) + mg \tag{12.31}$$

By elimination of force F in the aforementioned equations, we get

$$\begin{cases} M\ddot{x}_1 + m\ddot{y} = u \\ \ddot{y}\cos(\varphi) - \ddot{z}\sin(\varphi) = -g\sin(\varphi) \end{cases} \tag{12.32}$$

For a fixed cable length L:

$$z = L\cos(\varphi) \qquad\qquad y = x_1 + L\sin(\varphi)$$
$$\dot{z} = -L\dot{\varphi}\sin(\varphi) \qquad\qquad \dot{y} = \dot{x}_1 + L\dot{\varphi}\cos(\varphi) \tag{12.33}$$
$$\ddot{z} = -L\ddot{\varphi}\sin(\varphi) - L\dot{\varphi}^2\cos(\varphi) \qquad \ddot{y} = \ddot{x}_1 + L\ddot{\varphi}\cos(\varphi) - L\dot{\varphi}^2\sin(\varphi)$$

By substituting (12.33) into (12.32), we get

$$\begin{cases} (M+m)\ddot{x}_1 + mL(\ddot{\varphi}\cos(\varphi) - \dot{\varphi}^2\sin(\varphi)) = u \\ \ddot{x}_1\cos(\varphi) + L\ddot{\varphi} = -g\sin(\varphi) \end{cases} \tag{12.34}$$

We define the system's state vector to be

$$\begin{pmatrix} x_1 \\ x_2 \\ x_3 \\ x_4 \end{pmatrix} = \begin{pmatrix} x_1 \\ \dot{x}_1 \\ \varphi \\ \dot{\varphi} \end{pmatrix} \qquad (12.35)$$

Now, we can write down the nonlinear system state equations and perform a linearization.

Guidelines

You need to design the system with the parameters given below. The value of the parameters depends on a number of chosen constants, namely $U, V, W, X, Y,$ and Z. These constants are given as digits between 0 and 9. You could pick them randomly, or they might be provided to you.

Numeric values of the system's parameters are given in the following table:

Parameter	Value
Mass of the cart M	$100(5 + Z)$ [kg]
Mass load m	$100(1 + Y)$ [kg]
Cable length L	$5 + X$ [m]
Gravitation acceleration constant g	9.8 [N/kg]

For example, if $U = 4, V = 5, W = 6, X = 7, Y = 8,$ and $Z = 9$, then we get $M = 1400$ kg; $m = 900$ kg; $L = 12$ m.

Questions

1. Write down the nonlinear state-space equations using Equations (12.34) and the state vector defined in (12.35). Equations must be written in the form $\dot{x} = f(x, u)$.

2. Perform the linearization of nonlinear equations that are obtained around the equilibrium point $\{\varphi = 0; \dot{\varphi} = 0\}$ (hint: for small angles: $\cos(\varphi) \approx 1$; $\sin(\varphi) \approx \varphi$; $\sin^2(\varphi) \approx 0$; $\dot{\varphi}^2 \approx 0$). For the system's output y, prove that you get the system of the following shape:

$$\begin{cases} \dot{x} = Ax + Bu \\ y = Cx + Du \end{cases}$$

$$A = \begin{pmatrix} 0 & 1 & 0 & 0 \\ 0 & 0 & a_{23} & 0 \\ 0 & 0 & 0 & 1 \\ 0 & 0 & a_{43} & 0 \end{pmatrix}; \quad B = \begin{pmatrix} 0 \\ b_2 \\ 0 \\ b_4 \end{pmatrix}; \quad C = (c_1 \quad 0 \quad c_3 \quad 0); \quad D = 0$$

(Find all the constants in the matrix.)

3. We define the three possible outputs of the system:

   y: location of the load

   y_c: location of the cart ($y_c = x_1$)

   y_R: the cable angle relative to the cart ($y_R = \varphi$)

Compute three transfer functions between the input u and each of the outputs. Draw (approximately) the locations of the poles and zeros of each transfer function. What can be said about the stability of each function? Is the nonlinear system with the output y asymptotically stable? Explain.

4. Examine the stabilizability and detectability of each of the three aforementioned systems (with the given realization).

From now on, let's assume that the system has only y output.

5. Is it possible to stabilize the system using a fixed serial proportional controller (not state-space) K_p? Explain.

6. Implement and show the linear and nonlinear Simulink model of the system in an open loop, and present the response to the initial conditions

$$\begin{bmatrix} x_1 \\ x_2 \\ x_3 \\ x_4 \end{bmatrix}_0 = \begin{bmatrix} (W+1)[m] \\ 0 \\ 0.035(1+V)[rad] \\ 0 \end{bmatrix}$$ (with zero input) for both systems (all four of the

state variables, and a graph of the output y).

7. It is required to plan a controller that minimizes the following price function:

$$J = \int_0^\infty [x(t)^T Q x(t) + \rho u^2(t)] dt$$

Given access to all state variables, consider the state controller K for the following values of ρ and Q:

$$\rho \left[10^{-\left(2+\frac{U}{10}\right)}, 10^{-\left(5+\frac{U}{10}\right)}, 10^{-\left(10+\frac{U}{10}\right)} \right]$$

$$Q_1 = \begin{pmatrix} 0.001 & 0 & 0 & 0 \\ 0 & 0.001 & 0 & 0 \\ 0 & 0 & 1 & 0 \\ 0 & 0 & 0 & 0.001 \end{pmatrix}, \quad Q_2 = \begin{pmatrix} 1 & 0 & 0 & 0 \\ 0 & 0.001 & 0 & 0 \\ 0 & 0 & 0.001 & 0 \\ 0 & 0 & 0 & 0.001 \end{pmatrix}$$

(A total of six combinations.)

Note that the value of constant U was chosen at the beginning of this project.

a. Compute six optimal controllers for the matrices given above. How does selecting each matrix Q affect the cost function J?

b. Compute the closed-loop eigenvalues of the system in all six cases. Plot symmetric root locus as a function of ρ for each case.

c. Implement the closed-loop linear model with optimal controllers from part (a) and plot system's response to the initial conditions from Question 6. Plot separately the appropriate control effort u.

d. Repeat part (c) for the nonlinear system. Test for which initial conditions it stabilizes. Find the boundary initial conditions of the kind $(a,0,a,0)^T$ for which the nonlinear system is stabilizing.

8. Now, we assume that the system has uncertainty, and there are disturbances and noise affecting its response. The model is described in the following way:

$$\dot{x} = Ax + Bu + Gw$$

$$y = Cx + v$$

where G is the identity matrix, matrix C has dimensions 1×4, and w, v are white Gaussian independently distributed noises with zero mean:

$$w \sim N(0_{4 \times 1}, w_{var} \cdot I_{4 \times 4})$$

$$v \sim N(0_{2 \times 1}, v_{var})$$

where w_{var} and v_{var} are scalars.

In the first part of the question, we set $v_{var} = 10^{-7}$.

For w_{var} taking the values: $\left[10^{-\left(1+\frac{V}{10}\right)}, 10^{-\left(3+\frac{V}{10}\right)}\right]$, answer the following questions:

a. Are the necessary conditions for the Kalman filter satisfied?

b. Find the filter L_{Kalman} for each value of w_{var}. Specify the Kalman filter's eigenvalues for each case. Note that L_{Kalman} has dimensions 4×1, and you can use the MATLAB command "are" to compute it.
   Build the closed-loop model with noise state controller and the Kalman filter in Simulink. Pick one of the controllers from Question 7.

c. For each value w_{var}, find and show the system's response to the initial conditions from Question 6 for both observers $[L_{Kalman}, L_1]$ (total four combinations).

d. Is the system always stable in the four cases above? When it is stable, which of the two observers gives a better response?

e. The observer's cost function is defined as a quadratic error $J(t) = (x(t) - \hat{x}(t))^T (x(t) - \hat{x}(t))$. Compare the quadratic error of the Kalman filter with L_1 observer for one value of w_{var} for which the response is stable. Explain the differences between the errors obtained by each of the observers.
   Now, set $w_{var} = 10^{-7}$.

f. For v_{var} taking the following values: $\left[10^{-\left(1+\frac{W}{10}\right)}, 10^{-\left(3+\frac{W}{10}\right)}\right]$, repeat parts (b) to (e) above.

g. When is it beneficial to use the Kalman filter in this system? Explain.

Bonus Question

The purpose of this question is to design a system that moves a certain load from one place to another in the shortest time.

Only in this exercise we will assume that $M = 400$ kg, $m = 600$ kg, and $L = 4$ m. It is necessary to design an optimal controller and observer (not necessarily optimal) for the linear system, which will also work with the nonlinear system. The nonlinear system starts near the origin equilibrium point. Using the appropriate input, the nonlinear system (the hook) must be brought to position $y = -5$ [m] and then wait until the hook oscillations (swinging) are dampened enough (the hook does not swing more than ±15 [cm]). Then the system should move to $y = +5$ [m] and wait until swinging is small enough. You have to design the appropriate reference input signal (using the Signal Builder block of Simulink). Time is calculated from the moment the cart motion starts from the origin to the arrival of the load to $y = 5$ [m]. Additional limitations are $|u| < 500$; $|\dot{u}| < 500$; overshoot < 50%.

Write down your design process. It must be shown that all the restrictions and requirements are satisfied by showing plots of inputs, states, and outputs. The overshoot and settling time should be clearly outlined.

APPENDIX
Math Compendium

A   Notation and Nomenclature

$x(t), y(t)$: continuous-time signals	$\in$: element of (belongs to)
$x[n], y[n]$: discrete-time signals	$\mathbb{N}$: set of natural numbers: 0, 1, 2, 3, …
$X(\omega), X(z), X(s)$: transformed signals	$\mathbb{Z}$: set of integer numbers: $0, \pm 1, \pm 2, \ldots$
N, T: signal period (N_0, T_0: fundamental period)	$\mathbb{Q}$: set of rational numbers (fractions)
f: signal frequency (f_0: fundamental frequency)	$\mathbb{R}$: set of real numbers ($\mathbb{R}^+$ positive)
ω: angular frequency (ω_0: specific angular frequency)	$\mathbb{C}$: set of complex numbers
H, G: transfer function; system operator	$\forall$: for all (any)
T_s: sampling interval	$\exists$: exists
θ: signal phase	$\equiv$: identically equal
$\operatorname{sinc}(t) = \dfrac{\sin(t)}{t}$	$\operatorname{sinc}_\pi(t) = \dfrac{\sin(\pi t)}{\pi t}$
$\mathcal{U}(t) = \begin{cases} 1, & t \geq 0 \\ 0, & t < 0 \end{cases}$	$\operatorname{ramp}(t) = \begin{cases} t, & t \geq 0 \\ 0, & t < 0 \end{cases}$
$\operatorname{rect}(t) = \mathcal{U}(t+0.5) - \mathcal{U}(t-0.5) = \begin{cases} 1, & \|t\| \leq 0.5 \\ 0, & \|t\| > 0.5 \end{cases}$	$\operatorname{tri}(t) = \begin{cases} 1-\|t\|, & \|t\| < 1 \\ 0, & \|t\| \geq 1 \end{cases}$

B   Trigonometric Identities

1	$\cos(\theta_1 \pm \theta_2) = \cos(\theta_1)\cos(\theta_2) \mp \sin(\theta_1)\sin(\theta_2)$
2	$\sin(\theta_1 \pm \theta_2) = \sin(\theta_1)\cos(\theta_2) \pm \cos(\theta_1)\sin(\theta_2)$
3	$2\cos(\theta_1)\cos(\theta_2) = \cos(\theta_1 - \theta_2) + \cos(\theta_1 + \theta_2)$
4	$2\sin(\theta_1)\sin(\theta_2) = \cos(\theta_1 - \theta_2) - \cos(\theta_1 + \theta_2)$
5	$2\sin(\theta_1)\cos(\theta_2) = \sin(\theta_1 - \theta_2) + \sin(\theta_1 + \theta_2)$
6	$\cos\left(\theta \pm \dfrac{\pi}{2}\right) = \mp \sin(\theta)$
7	$\sin\left(\theta \pm \dfrac{\pi}{2}\right) = \pm \cos(\theta)$

8	$\cos(2\theta) = \cos^2(\theta) - \sin^2(\theta)$
9	$\sin(2\theta) = 2\sin(\theta)\cos(\theta)$
10	$2\cos^2(\theta) = 1 + \cos(2\theta)$
11	$2\sin^2(\theta) = 1 - \cos(2\theta)$
12	$4\cos^3(\theta) = 3\cos(\theta) + \cos(3\theta)$
13	$4\sin^3(\theta) = 3\sin(\theta) - \sin(3\theta)$
14	$\cos^2(\theta) + \sin^2(\theta) = 1$

C Complex Numbers

$j = \sqrt{-1}; \quad j^2 = -1; \quad j^3 = -j; \quad j^4 = 1; \quad j^5 = j; \ldots$

Cartesian: $s = a + bj$	Polar: $s = Re^{j\theta}$
Real value: $Re\{s\} = a$	Real value: $Re\{s\} = R\cos\theta$
Imaginary value: $Im\{s\} = b$	Imaginary value: $Im\{s\} = R\sin\theta$
Absolute value (magnitude): $|s| = \sqrt{a^2 + b^2}$	Absolute value (magnitude): $|s| = R$
Angle: $\angle s = \tan^{-1}\left(\dfrac{b}{a}\right)$	Angle: $\angle s = \theta$
Conjugate: $\bar{s} = a - bj$	Conjugate: $\bar{s} = Re^{-j\theta}$

C.1 Euler Theorem and de Moivre's Formula

$e^{\pm j\theta} = \cos\theta \pm j\sin\theta$	$\cos\theta = \dfrac{e^{j\theta} + e^{-j\theta}}{2}$	$\sin\theta = \dfrac{e^{j\theta} - e^{-j\theta}}{2j}$
$|e^{\pm j\theta}| = 1$	$\angle e^{\pm j\theta} = \pm\theta$	$(re^{j\theta})^{\frac{1}{n}} = r^{1/n}e^{\frac{j(\theta + 2\pi k)}{n}}; \quad n = \mathbb{N}, k = 0, \ldots, n-1$
$e^{\pm j\pi/2} = \pm j$	$e^{\pm j 2n\pi} = 1; \quad n \in \mathbb{N}$	$e^{\pm j(2n+1)\pi} = -1; \quad n \in \mathbb{N}$

C.2 Multiplication

$$(a + bj)(c + dj) = (ac - bd) + (ad + bc)j$$

$$R_1 e^{j\theta_1} R_2 e^{j\theta_2} = R_1 R_2 e^{j(\theta_1 + \theta_2)}$$

C.3 Division

$$\frac{a + bj}{c + dj} = \frac{ac + bd}{c^2 + d^2} + \frac{bc - ad}{c^2 + d^2} j$$

$$R_1 e^{j\theta_1} / R_2 e^{j\theta_2} = \frac{R_1}{R_2} e^{j(\theta_1 - \theta_2)}$$

D Algebra

D.1 Inequalities

- Triangle inequality: $\forall a, b \in \mathbb{R}: |a + b| \leq |a| + |b|$

  This inequality could be generalized to any finite number of a_i:

$$\left|\sum_{i=1}^{n} a_i\right| \leq \sum_{i=1}^{n} |a_i|$$

The equality reached only when a_i are all positive or all negative.

- $\forall a, b \in \mathbb{R}: |a-b| \geq ||a|-|b||$.
- $\forall a, b \in \mathbb{R}: a^2 + b^2 \geq 2|ab|$ (in particular from $(a-b)^2 \geq 0$ follows $a^2 + b^2 \geq 2ab$). The equality is reached if and only if $|a|=|b|$.
- $\forall a, b \in \mathbb{R}$ of the same sign ($ab > 0$):

$$\frac{a}{b} + \frac{b}{a} \geq 2$$

  and the equality is reached if and only if $a = b$.

- Cauchy inequality for finite number of $a_i \geq 0$:

$$\frac{a_1 + a_2 + \cdots + a_n}{n} = \frac{1}{n}\sum_{i=1}^{n} a_i \geq \sqrt[n]{\prod_{i=1}^{n} a_i} = \sqrt[n]{a_1 a_2 \cdots a_n}$$

  which means that the geometric mean is not greater than the arithmetic mean.

- Hölder's inequality: $\forall a_i, b_i \in \mathbb{R}, \forall p > 1$:

$$\left|\sum_{i=1}^{n} a_i b_i\right| \leq \left(\sum_{i=1}^{n} |a_i|^p\right)^{\frac{1}{p}} \left(\sum_{i=1}^{n} |a_i|^{\frac{p}{p-1}}\right)^{\frac{p-1}{p}}$$

D.2 Polynomials

Polynomial of the free variable s is the sum of powers of s multiplied by constant coefficients:

$$P(s) = \sum_{k=0}^{n} a_k s^k = a_n s^n + a_{n-1} s^{n-1} + \cdots + a_1 s + a_0$$

where a_k ($k = 0, 1, \ldots, n$) are constant coefficients and n is the order of the polynomial.

Theorem

Any polynomial of the order n has exactly n complex roots and it could be factorized as follows:

$$a_n s^n + a_{n-1} s^{n-1} + \cdots + a_1 s + a_0 = a_n (s - s_1)(s - s_2) \cdots (s - s_n),$$

where $s_1, \ldots, s_n$ are the roots.

Example

$$2s^2 + 6s + 4 = 2(s+2)(s+1)$$

where -1 and -2 are the roots of the polynomial $2s^2 + 6s + 4$.

Appendix

Vieta's Formulas

For a polynomial $P(s) = a_n s^n + a_{n-1} s^{n-1} + \cdots + a_1 s + a_0$,

$$\begin{cases} s_1 + s_2 + \cdots + s_n = -\dfrac{a_{n-1}}{a_n} \\ (s_1 s_2 + s_1 s_3 + s_1 s_n) + (s_2 s_3 + s_2 s_4 + \cdots + s_2 s_n) + \cdots + s_{n-1} s_n = \dfrac{a_{n-2}}{a_n} \\ \vdots \\ s_1 s_2 \ldots s_n = (-1)^n \dfrac{a_0}{a_n} \end{cases}$$

In other words, the sum and product of roots could be derived from the coefficients without computing the roots.

Example

The sum of roots of $2s^2 + 6s + 4 = 0$ is $-\dfrac{6}{2} = -3$ and the product (multiplication) is $\dfrac{4}{2} = 2$.

E Calculus

E.1 Derivative Rules

Chain rule	$\dot{f}(g(t)) = \dfrac{d}{dt} f(g(t)) = \dfrac{df(g)}{dg} \dot{g}(t)$
Linearity	$\dfrac{d}{dt}(af(t) + bg(t)) = a\dot{f}(t) + b\dot{g}(t)$
Product rule	$\dfrac{d}{dt}(f(t)g(t)) = \dot{f}(t)g(t) + f(t)\dot{g}(t)$
Quotient rule	$\dfrac{d}{dt}\left(\dfrac{f(t)}{g(t)}\right) = \dfrac{\dot{f}(t)g(t) - f(t)\dot{g}(t)}{g^2(t)}$
Leibniz's rule	$\dfrac{d}{dt}\left[\int_{a(t)}^{b(t)} f(\lambda, t)\, d\lambda\right] = f(b(t), t)\dfrac{db(t)}{dt} - f(a(t), t)\dfrac{da(t)}{dt} + \int_{a(t)}^{b(t)} \dfrac{\partial f(\lambda, t)}{\partial t}\, d\lambda$

E.2 Derivatives Table

1	$\dfrac{d}{dt}(t^n) = nt^{n-1}$		4	$\dfrac{d}{dt}(\sin at) = a\cos at$
2	$\dfrac{d}{dt}(e^{at}) = ae^{at}$		5	$\dfrac{d}{dt}(\cos at) = -a\sin at$
3	$\dfrac{d}{dt}(\ln t) = \dfrac{1}{t}$		6	$\dfrac{d}{dt}(a^t) = a^t \ln a$

Math Compendium

E.3 Integration

Linearity	$\int af(t) + bg(t)\,dt = a\int f(t)\,dt + b\int g(t)\,dt$
Integration by parts	$\int_a^b f(t)\dot{g}(t)\,dt = f(t)g(t)\big|_a^b - \int_a^b \dot{f}(t)g(t)\,dt$
Change in variable	$\lambda = \varphi(t): \int_a^b f(\lambda)\,d\lambda = \int_{\varphi^{-1}(a)}^{\varphi^{-1}(b)} f(\varphi(t))\dot{\varphi}(t)\,dt$

E.4 Indefinite Integrals

1	$\int t^n\,dt = \dfrac{t^{n+1}}{n+1}; \quad n \neq -1$
2	$\int \dfrac{1}{t}\,dt = \ln|t|$
3	$\int a^t\,dt = \dfrac{a^t}{\ln a}$
4	$\int e^{at}\,dt = \dfrac{e^{at}}{a}$
5	$\int t e^{at}\,dt = e^{at}\left(\dfrac{t}{a} - \dfrac{1}{a^2}\right)$
6	$\int \cos(at)\,dt = \dfrac{1}{a}\sin(at)$
7	$\int \sin(at)\,dt = -\dfrac{1}{a}\cos(at)$
8	$\int t\cos(at)\,dt = \dfrac{1}{a^2}(\cos(at) + at\sin(at))$
9	$\int t\sin(at)\,dt = \dfrac{1}{a^2}(\sin(at) - at\cos(at))$
10	$\int t^n \cos t\,dt = x^n \sin t - n\int t^{n-1}\sin t\,dt$
11	$\int t^n \sin t\,dt = -t^n\cos t + n\int t^{n-1}\cos t\,dt$
12	$\int t^n e^{at}\,dt = \dfrac{1}{a}\left[t^n e^{at} - n\int t^{n-1}e^{ax}\,dt\right]$
13	$\int e^{at}\sin bt\,dt = \dfrac{e^{at}}{a^2+b^2}(a\sin bt - b\cos bt)$
14	$\int e^{at}\cos bt\,dt = \dfrac{e^{at}}{a^2+b^2}(a\cos bt + b\sin bt)$
15	$\int \cos(at)\cos(bt)\,dt = \dfrac{\sin[(a-b)t]}{2(a-b)} + \dfrac{\sin[(a+b)t]}{2(a+b)}$
16	$\int \sin(at)\sin(bt)\,dt = \dfrac{\sin[(a-b)t]}{2(a-b)} - \dfrac{\sin[(a+b)t]}{2(a+b)}$
17	$\int \sin(at)\cos(bt)\,dt = -\dfrac{\cos[(a-b)t]}{2(a-b)} - \dfrac{\cos[(a+b)t]}{2(a+b)}$
18	$\int \cos^2(at)\,dt = \dfrac{1}{4a}[2at + \sin(2at)]$

E.5 Definite Integrals

1	$\int_0^\infty \dfrac{t^{m-1}}{t^n+1}\,dt = \dfrac{\pi/n}{\sin(m\pi/n)}$
2	$\int_0^\infty t^n e^{-at}\,dt = \dfrac{n!}{a^{n+1}}; \quad a>0$
3	$\int_0^\infty e^{-at}\cos bt\,dt = \dfrac{a}{a^2+b^2}; \quad a>0$
4	$\int_0^\infty e^{-at}\sin bt\,dt = \dfrac{b}{a^2+b^2}; \quad a>0$
5	$\int_0^\infty e^{-a^2 t^2}\cos bt\,dt = \dfrac{\sqrt{\pi}}{2a}e^{-b^2/(4a^2)}; \quad a>0$
6	$\int_0^\infty \dfrac{\sin at}{t}\,dt = \dfrac{\pi}{2}\operatorname{sign}(a)$
7	$\int_0^\infty \left(\dfrac{\sin t}{t}\right)^2\,dt = \dfrac{\pi}{2}$
8	$\int_{-\infty}^\infty e^{\pm j2\pi at}\,dt = \delta(a)$

E.6 Finite Series

1	$\sum_{n=1}^{N} n = \dfrac{N(N+1)}{2}$		4	$\sum_{n=0}^{N} a^n = \dfrac{a^{N+1}-1}{a-1}; \quad a \neq 1$	
2	$\sum_{n=1}^{N} n^2 = \dfrac{N(N+1)(2N+1)}{6}$		5	$\sum_{n=0}^{N} \dfrac{N!}{(N-n)!n!} a^n b^{N-n} = (a+b)^N$	
3	$\sum_{n=1}^{N} n^3 = \dfrac{N^2(N+1)^2}{4}$		6	$\sum_{n=0}^{N} na^n = \dfrac{1-(N+1)a^N + Na^{N+1}}{(1-a)^2}$	

E.7 Infinite Series

Taylor Series

$$f(t) = \sum_{n=0}^{\infty}\left(\frac{1}{n!}\frac{d^n}{dt^n}f(a)\right)(t-a)^n = f(a) + \dot{f}(a)(t-a) + \frac{1}{2}\ddot{f}(a)(t-a)^2 + \frac{1}{6}\dddot{f}(a)(t-a)^3 + \cdots$$

$$e^t = \sum_{n=0}^{\infty}\frac{t^n}{n!} = 1 + t + \frac{t^2}{2} + \frac{t^3}{6} + \frac{t^4}{24} + \cdots$$

$$te^t = \sum_{n=0}^{\infty}\frac{t^n}{(n-1)!} = t + t^2 + \frac{t^3}{2} + \frac{t^4}{6} + \cdots$$

$$\frac{1}{1-t} = \sum_{n=0}^{\infty} t^n = 1 + t + t^2 + t^3 + t^4 + \cdots; \quad |t| < 1$$

$$\ln(1+t) = \sum_{n=1}^{\infty}(-1)^{n+1}\frac{t^n}{n} = t - \frac{t^2}{2} + \frac{t^3}{3} - \frac{t^4}{4} + \cdots$$

$$\cos t = \sum_{n=0}^{\infty}(-1)^n\frac{t^{2n}}{(2n)!} = 1 - \frac{t^2}{2} + \frac{t^4}{24} - \frac{t^6}{720} + \cdots$$

$$\sin t = \sum_{n=0}^{\infty}(-1)^n\frac{t^{2n+1}}{(2n+1)!} = t - \frac{t^3}{6} + \frac{t^5}{120} - \frac{t^7}{5040} + \cdots$$

Multivariable Taylor Series

A multivariable function $f(t)$ around the point a is approximated by

$$f(t+a) = f(a) + (\nabla f(a))^T t + \frac{1}{2} t^T \nabla^2 f(a) t + \text{(higher-order terms)}$$

where both a and t are vectors, ∇ denotes the gradient (vector of partial derivatives of f), and ∇^2 denotes the Hessian matrix of second-order derivatives.

F Signals and Systems

F.1 Linear Time Invariant Systems

Let's assume that the response of a system to input $u_1(t)$ is $y_1(t)$ and the response to another input $u_2(t)$ is $y_2(t)$. The *system is linear* if for any u_1 and u_2 and for any scalars α and β, the response of this system to $\alpha u_1(t) + \beta u_2(t)$ is $\alpha y_1(t) + \beta y_2(t)$. In other words, linearity is when you multiply the input by some constant, the output is multiplied by the same constant, and the response to sum of signals is the sum of responses.

A system is *time invariant* if for any scalar τ the response of this system to $u_1(t-\tau)$ is $y_1(t-\tau)$. In other words, time invariance means that the response to a shifted signal is shifted response (by the same time).

F.2 Fourier Series

Trigonometric Fourier series for a function $x(t)$ defined on interval $[a, a+T]$:

$$x(t) = \frac{a_0}{2} + \sum_{k=1}^{\infty} \left[a_k \cos\left(k\frac{2\pi}{T}t\right) + b_k \sin\left(k\frac{2\pi}{T}t\right) \right]$$

where $a_k, b_k \in \mathbb{R}$ are the *Fourier coefficients*,

$$a_k = \frac{2}{T} \int_a^{a+T} x(t) \cos\left(k\frac{2\pi}{T}t\right) dt; \qquad k = 0, 1, 2, \ldots$$

$$b_k = \frac{2}{T} \int_a^{a+T} x(t) \sin\left(k\frac{2\pi}{T}t\right) dt; \qquad k = 1, 2, \ldots$$

Complex Fourier series for a function $x(t)$ defined on interval $[a, a+T]$:

$$x(t) = \sum_{k=-\infty}^{\infty} C_k e^{jk\frac{2\pi}{T}t}$$

where $C_k \in \mathbb{C}$ are the *complex Fourier coefficients*:

$$C_k = \frac{1}{T} \int_a^{a+T} x(t) e^{jk\frac{2\pi}{T}t} dt; \qquad k = 0, \pm 1, \pm 2, \ldots$$

Relation between coefficients:

$$C_k = \begin{cases} 0.5(a_k - jb_k), & k = 1, 2, 3, \ldots \\ 0.5 a_0, & k = 0 \\ 0.5(a_{-k} + jb_{-k}), & k = -1, -2, -3, \ldots \end{cases}$$

Amplitude spectrum: $|C_k| = 0.5\sqrt{a_k^2 + b_k^2}$

Phase spectrum: $\angle C_k = \tan^{-1}\left(\frac{b_k}{a_k}\right)$

Power spectrum: $|C_k|^2 = 0.25(a_k^2 + b_k^2)$

Parseval's theorem: $\frac{1}{T}\int_a^{a+T} |x(t)|^2 dt = \sum_{k=-\infty}^{\infty} |C_k|^2$

F.3 Discrete Fourier Transform (DFT)

$$C_k = \sum_{n=0}^{N-1} x[n] e^{-j\frac{2\pi}{N}kn}; \qquad k = 0, 1, 2, \ldots, N-1$$

Inverse DFT is defined as

$$x[n] = \frac{1}{N} \sum_{k=0}^{N-1} C_k e^{j\frac{2\pi}{N}kn}; \qquad n = 0, 1, 2, \ldots, N-1$$

F.4 Convolution Integral

$$y(t) = x(t) * h(t) = \int_{-\infty}^{\infty} x(\tau) h(t-\tau) d\tau$$

F.5 Convolution Sum

$$y[n] = x[n] * h[n] = \sum_{k=-\infty}^{\infty} x[k] h[n-k]$$

Commutativity of convolution integral and sum:

$$x(t) * h(t) = h(t) * x(t)$$
$$x[n] * h[n] = h[n] * x[n]$$

F.6 Generalized Functions

Dirac Delta

Definition: $\delta(t) = \begin{cases} \infty, & t = 0 \\ 0, & t \neq 0 \end{cases}$ and $\int_{-\infty}^{\infty} \delta(t) dt = 1$.

1	$\delta(-t) = \delta(t)$		5	$\int_{-\infty}^{\infty} x(t) \delta(t-\tau) dt = x(\tau)$
2	$\delta(at) = \frac{\delta(t)}{\|a\|}$		6	$x(t) * \delta(t-\tau) = x(t-\tau)$
3	$\delta(t-\tau) = \begin{cases} \infty, & t = \tau \\ 0, & t \neq \tau \end{cases}$		7	$x(t) * \dot{\delta}(t) = \dot{x}(t)$
4	$x(t) \delta(t-\tau) = x(\tau) \delta(t-\tau)$		8	$\delta(t) = \frac{d}{dt} \mathcal{U}(t)$

Kronecker Delta

Definition: $\delta[n] = \begin{cases} 1, & n = 0 \\ 0, & n \neq 0 \end{cases}$.

1	$\delta[-n] = \delta[n]$		4	$\sum_{k=-\infty}^{\infty} \delta[n-k] a[k] = a[n]$
2	$\delta[an] = \delta[n]; \quad a \in \mathbb{Z} \setminus 0$		5	$\sum_{k=-\infty}^{\infty} \delta[n-k] \delta[k-j] = \delta[n-j]$
3	$\delta[n-k] = \begin{cases} 1, & n = k \\ 0, & n \neq k \end{cases}$			

F.7 Signal Energy and Power

Energy of $x(t)$: $E_\infty = \int_{-\infty}^{\infty} |x(t)|^2 dt$	Energy of $x[n]$: $E_\infty = \sum_{n=-\infty}^{\infty} |x[n]|^2$
Average power of $x(t)$: $$P_\infty = \lim_{T \to \infty} \frac{1}{2T} \int_{-T}^{T} |x(t)|^2 dt$$	Average power of $x[n]$: $$P_\infty = \lim_{N \to \infty} \frac{1}{2N+1} \sum_{n=-N}^{N} |x[n]|^2$$
Average power of T-periodic signal ($\forall a$): $$P_T = \frac{1}{T} \int_{a}^{a+T} |x(t)|^2 dt$$	Average power of N-periodic signal ($\forall a$): $$P_N = \frac{1}{N} \sum_{n=a+1}^{a+N} |x[n]|^2$$

F.8 Fourier Transform (FT)

FT Definition

Fourier transform $X(\omega) = \mathcal{F}\{x(t)\}$	Inverse Fourier transform $x(t) = \mathcal{F}^{-1}\{X(\omega)\}$
$X(\omega) = \int_{-\infty}^{\infty} x(t) e^{-j\omega t} dt$, $-\infty < \omega < \infty$	$x(t) = \frac{1}{2\pi} \int_{-\infty}^{\infty} X(\omega) e^{j\omega t} d\omega$

FT Properties

	Property Name	Signal	Transform
1	Linearity	$ax(t) + by(t)$	$aX(\omega) + bY(\omega)$
2	Time shift	$x(t-a); \quad a \in \mathbb{R}$	$X(\omega) e^{-j\omega a}$
3	Time scaling	$x(at); \quad a \neq 0$	$\frac{1}{|a|} X\left(\frac{\omega}{a}\right)$
4	Time reversal	$x(-t)$	$X(-\omega)$
5	Multiply by t^n	$t^n x(t); \quad n = 1, 2, \ldots$	$j^n \frac{d^n}{d\omega^n} X(\omega)$
6	Multiply by $e^{j\omega_0 t}$	$x(t) e^{j\omega_0 t}; \quad \omega_0 \in \mathbb{R}$	$X(\omega - \omega_0)$
7	Multiply by $\cos \omega_0 t$	$x(t) \cos \omega_0 t$	$\frac{1}{2}[X(\omega + \omega_0) + X(\omega - \omega_0)]$
8	Multiply by $\sin \omega_0 t$	$x(t) \sin \omega_0 t$	$\frac{j}{2}[X(\omega + \omega_0) - X(\omega - \omega_0)]$
9	Derivative	$\frac{d^n}{dt^n} x(t)$	$(j\omega)^n X(\omega)$
10	Integral	$\int_{-\infty}^{t} x(\tau) d\tau$	$\frac{1}{j\omega} X(\omega) + \pi X(0) \delta(\omega)$
11	Convolution	$x(t) * y(t)$	$X(\omega) Y(\omega)$
12	Multiplication	$x(t) y(t)$	$\frac{1}{2\pi} X(\omega) * Y(\omega)$
13	Duality	$X(t)$	$2\pi x(-\omega)$
14	Parseval's theorem	$\int_{-\infty}^{\infty} |x(t)|^2 dt = \frac{1}{2\pi} \int_{-\infty}^{\infty} |X(\omega)|^2 d\omega$	

FT Common Pairs

#		
1	$\delta(t)$	1
2	$\delta(t-a); \quad a \in \mathbb{R}$	$e^{-j\omega a}$
3	$\mathcal{U}(t)$	$\pi\delta(\omega) + \dfrac{1}{j\omega}$
4	$\mathcal{U}(t) - 0.5$	$\dfrac{1}{j\omega}$
5	$e^{-at}\mathcal{U}(t); \quad a > 0$	$\dfrac{1}{a+j\omega}$
6	1	$2\pi\delta(\omega)$
7	$e^{j\omega_0 t}; \quad \omega_0 \in \mathbb{R}$	$2\pi\delta(\omega - \omega_0)$
8	$\cos(\omega_0 t)$	$\pi[\delta(\omega+\omega_0) + \delta(\omega-\omega_0)]$
9	$\sin(\omega_0 t)$	$j\pi[\delta(\omega+\omega_0) - \delta(\omega-\omega_0)]$
10	$\cos(\omega_0 t + \theta)$	$\pi[e^{-j\theta}\delta(\omega+\omega_0) + e^{j\theta}\delta(\omega-\omega_0)]$
11	$\sin(\omega_0 t + \theta)$	$j\pi[e^{-j\theta}\delta(\omega+\omega_0) - e^{j\theta}\delta(\omega-\omega_0)]$
12	$e^{-a\|t\|}$	$\dfrac{2a}{a^2 + \omega^2}$
13	$\mathcal{U}(t+a) - \mathcal{U}(t-a)$	$\dfrac{2\sin(a\omega)}{\omega}$
14	$\dfrac{\sin(\omega_0 t)}{\pi t}$	$\mathcal{U}(\omega+\omega_0) - \mathcal{U}(\omega-\omega_0)$
15	$\text{rect}(t)$	$\text{sinc}_\pi\left(\dfrac{\omega}{2\pi}\right)$
16	$\text{sinc}_\pi(t)$	$\text{rect}\left(\dfrac{\omega}{2\pi}\right)$
17	$\text{tri}(t)$	$\left(\text{sinc}_\pi\left(\dfrac{\omega}{2\pi}\right)\right)^2$
18	$\left(\text{sinc}_\pi\left(\dfrac{t}{2\pi}\right)\right)^2$	$\text{tri}\left(\dfrac{\omega}{2\pi}\right)$

F.9 Partial Fractions Expansion

Given a general rational transfer function with real distinct poles, repeated poles, and complex poles:

$$H(s) = \frac{N(s)}{(s+p_1)\cdots(s+p_n)(s+p_m)^r(s^2+as+b)} = \frac{K_1}{s+p_1} + \cdots + \frac{K_n}{s+p_n} +$$

$$+ \frac{K_{m1}}{(s+p_m)^r} + \frac{K_{m2}}{(s+p_m)^{r-1}} + \cdots + \frac{K_{mr}}{s+p_m} +$$

$$+ \frac{K_l s + K_{l+1}}{s^2+as+b}$$

$$K_i = \lim_{s \to -p_i}(s+p_i)H(s); \quad i = 1, \ldots, n$$

$$K_{mi} = \lim_{s \to -p_m}\left(\frac{1}{(i-1)!}\frac{d^{i-1}}{ds^{i-1}}((s+p_m)^r H(s))\right); \quad i = 1, \ldots, r$$

F.10 Laplace Transform (LT)

LT Definition
Assume that $\forall (t<0): f(t) = 0$.

Laplace transform $F(s) = \mathcal{L}\{f(t)\}$	Inverse Laplace transform $f(t) = \mathcal{L}^{-1}\{F(s)\}$
$F(s) = \int_{0-}^{\infty} f(t)e^{-st}\,dt$	$f(t) = \dfrac{1}{2\pi j} \int_{c-j\infty}^{c+j\infty} F(s)e^{st}\,ds$

LT Properties

	Property Name	Signal	Transform
1	Linearity	$af_1(t) + bf_2(t)$	$aF_1(s) + bF_2(s)$
2	Right shift in time	$f(t-a)\mathcal{U}(t-a);\quad a>0$	$F(s)e^{-as}$
3	Time scaling	$f(at);\quad a>0$	$\dfrac{1}{a}F\left(\dfrac{s}{a}\right)$
4	Multiply by t^n	$t^n f(t);\quad n=1,2,\ldots$	$(-1)^n \dfrac{d^n}{ds^n} F(s)$
5	Multiply by e^{at}	$f(t)e^{at};\quad a\in\mathbb{C}$	$F(s-a)$
6	Multiply by $\cos\omega_0 t$	$f(t)\cos\omega_0 t$	$\dfrac{1}{2}[F(s+j\omega_0) + F(s-j\omega_0)]$
7	Multiply by $\sin\omega_0 t$	$f(t)\sin\omega_0 t$	$\dfrac{j}{2}[F(s+j\omega_0) - F(s-j\omega_0)]$
8	Derivative	$\dot{f}(t)$	$sF(s) - f(0)$
9	Second derivative	$\ddot{f}(t)$	$s^2 F(s) - sf(0) - \dot{f}(0)$
10	nth derivative	$f^{(n)}(t)$	$s^n F(s) - \sum_{k=1}^{n} s^{n-k} f^{(k-1)}(0)$
11	Integral	$\int_0^t f(\tau)\,d\tau$	$\dfrac{1}{s}F(s)$
12	Convolution	$f_1(t) * f_2(t)$	$F_1(s)F_2(s)$
13	Initial-value theorem	\multicolumn{2}{c}{$f(0) = \lim_{s\to\infty}(sF(s))$}
14	Final-value theorem	\multicolumn{2}{c}{If $\lim_{t\to\infty} f(t)$ exists, then $\lim_{t\to\infty} f(t) = \lim_{s\to 0}(sF(s))$}

Appendix

LT Common Pairs

1	$\delta(t)$	1
2	$\delta(t-a); \quad a > 0$	e^{-as}
3	$\mathcal{U}(t)$	$\dfrac{1}{s}$
4	$\mathcal{U}(t) - \mathcal{U}(t-a); \quad a > 0$	$\dfrac{1 - e^{-as}}{s}$
5	$e^{-at}\mathcal{U}(t); \quad a \in \mathbb{C}$	$\dfrac{1}{s+a}$
6	$t^n e^{-at}\mathcal{U}(t); \quad n = 1, 2, \ldots$	$\dfrac{n!}{(s+a)^{n+1}}$
7	$(1 - at)e^{-at}\mathcal{U}(t)$	$\dfrac{s}{(s+a)^2}$
8	$\cos(at)\,\mathcal{U}(t)$	$\dfrac{s}{s^2 + a^2}$
9	$\sin(at)\,\mathcal{U}(t)$	$\dfrac{a}{s^2 + a^2}$
10	$\dfrac{-e^{-at} + e^{-bt}}{a - b}\mathcal{U}(t)$	$\dfrac{1}{(s+a)(s+b)}$
11	$\dfrac{ae^{-at} - be^{-bt}}{a - b}\mathcal{U}(t)$	$\dfrac{s}{(s+a)(s+b)}$
12	$\dfrac{(c-b)e^{-at} + (a-c)e^{-bt} + (b-a)e^{-ct}}{(a-b)(b-c)(c-a)}\mathcal{U}(t)$	$\dfrac{1}{(s+a)(s+b)(s+c)}$
13	$\dfrac{-a(c-b)e^{-at} - b(a-c)e^{-bt} - c(b-a)e^{-ct}}{(a-b)(b-c)(c-a)}\mathcal{U}(t)$	$\dfrac{s}{(s+a)(s+b)(s+c)}$
14	$\dfrac{-a(c-b)e^{-at} - b(a-c)e^{-bt} - c(b-a)e^{-ct}}{(a-b)(b-c)(c-a)}\mathcal{U}(t)$	$\dfrac{s^2}{(s+a)(s+b)(s+c)}$
15	$\dfrac{e^{-at} - e^{-bt} + (a-b)te^{-bt}}{(a-b)^2}\mathcal{U}(t)$	$\dfrac{1}{(s+a)(s+b)^2}$
16	$\dfrac{-ae^{-at} + ae^{-bt} - b(a-b)te^{-bt}}{(a-b)^2}\mathcal{U}(t)$	$\dfrac{s}{(s+a)(s+b)^2}$
17	$\dfrac{a^2 e^{-at} - b(2a-b)e^{-bt} + b^2(a-b)te^{-bt}}{(a-b)^2}\mathcal{U}(t)$	$\dfrac{s^2}{(s+a)(s+b)^2}$
18	$e^{-bt}\cos(at)\,\mathcal{U}(t)$	$\dfrac{s+b}{(s+b)^2 + a^2}$
19	$e^{-bt}\sin(at)\,\mathcal{U}(t)$	$\dfrac{a}{(s+b)^2 + a^2}$

F.11 Ordinary Differential Equations

Let $y^{(n)} + a_{n-1} y^{(n-1)} + \cdots + a_1 y^{(1)} + a_0 y = f(t)$ be the ordinary differential equation (ODE), where $f(t)$ is some function of time.

We apply the Laplace transform to both sides of the equation $\dot{y}(t) \xrightarrow{\text{Laplace}} sY(s) - y(0)$ and get an algebraic equation. We then solve this algebraic equation with regard to s to find $Y(s)$. Finally, we apply the inverse Laplace transform to get the solution $y(t)$. Similarly, we can solve discrete-time difference equations using the Z-transform.

A general solution of a linear homogeneous equation [where $f(t) = 0$] is $y(t) = C_1 e^{s_1 t} + C_2 e^{s_2 t} + \cdots + C_n e^{s_n t}$, where $C_1, C_2, \ldots, C_n$ are constants, and $s_1, s_2, \ldots, s_n$ are the solutions of the polynomial equation $s^n + a_{n-1} s^{n-1} + \cdots + a_1 s + a_0 = 0$. The constants C_i are found from the initial conditions of the ODE.

The general solution of a first-order ODE: $\frac{\partial x(t)}{\partial t} + p(t)x(t) = q(t); x(t_0) = x_0$ is

$$x(t) = \exp\left(-\int_{t_0}^{t} p(\tau)d\tau\right)\left(x_0 + \int_{t_0}^{t}\left[q(\tau)\exp\left(\int_{t_0}^{\tau} p(\sigma)d\sigma\right)\right]d\tau\right)$$

F.12 Z-Transform

Z-Transform Definition

Assume that $\forall (n < 0): x[n] = 0$.

Laplace transform $X(z) = Z\{x(t)\}$	Inverse Laplace transform $x[n] = Z^{-1}\{X(z)\}$
$X(z) = \sum_{n=0}^{\infty} x[n] z^{-n}$	$x[n] = \frac{1}{2\pi j} \int_{\text{convergence circle}} X(z) z^{k-1} dz$

Z-Transform Properties

	Property Name	Signal	Transform
1	Linearity	$ax[n] + by[n]$	$aX(z) + bY(z)$
2	Right shift in time	$x[n-a];\ a \in \mathbb{N}$	$z^{-a} X(z)$
3	Left shift in time	$x[n+a];\ a \in \mathbb{N}$	$z^a X(z) - \sum_{k=0}^{a-1} x[k] z^{a-k}$
4	Multiply by n	$nx[n]$	$-z \frac{d}{dz} X(z)$
5	Multiply by n^2	$n^2 x[n]$	$z \frac{d}{dz} X(z) + z^2 \frac{d^2}{dz^2} X(z)$
6	Multiply by a^n	$a^n x[n]$	$X\left(\frac{z}{a}\right)$
7	Multiply by $\cos \omega_0 n$	$x[n] \cos \omega_0 n$	$\frac{1}{2}[X(e^{j\omega_0} z) + X(e^{j\omega_0} z)]$
8	Multiply by $\sin \omega_0 n$	$x[n] \sin \omega_0 n$	$\frac{j}{2}[X(e^{j\omega_0} z) - X(e^{j\omega_0} z)]$
9	Summation	$\sum_{i=0}^{n} x[i]$	$\frac{z}{z-1} X(z)$
10	Convolution	$x[n] * y[n]$	$X(z) Y(z)$
11	Initial-value theorem		$x[0] = \lim_{z \to \infty}(X(z))$
12	Final-value theorem		If $X(z)$ is rational and the poles of $(z-1)X(z)$ have magnitudes <1, then $\lim_{n \to \infty} x[n] = [(z-1)X(z)]_{z=1}$

Z-Transform Common Pairs

1	$\delta[n]$	1
2	$\delta[n-a];\quad a=1,2,\ldots$	z^{-q}
3	$\mathcal{U}[n]$	$\dfrac{z}{z-1}$
4	$\mathcal{U}[n]-\mathcal{U}[n-a];\quad a=1,2,\ldots$	$\dfrac{z^a-1}{z^{a-1}(z-1)}$
5	$a^{n-k}\,\mathcal{U}[n-k];\quad a\in\mathbb{C},\,k=0,1,2,\ldots$	$\dfrac{z^{1-k}}{z-a}$
6	$na^n\,\mathcal{U}[n]$	$\dfrac{az}{(z-a)^2}$
7	$(n+1)\mathcal{U}[n]$	$\dfrac{z^2}{(z-1)^2}$
8	$n^2 a^n \mathcal{U}[n]$	$\dfrac{az(z+a)}{(z-a)^3}$
9	$\dfrac{a^{n+k}-b^{n+k}}{a-b}\mathcal{U}[n+k-1];\quad k\in\mathbb{Z}$	$\dfrac{z^{k+1}}{(z-a)(z-b)}$
10	$\left(\dfrac{(a-b)(n+k)a^{n+k-1}-a^{n+k}+b^{n+k}}{(a-b)^2}\right);\quad k=2,1,0,\ldots$	$\dfrac{z^{k+1}}{(z-a)^2(z-b)}$
11	$\cos[bn]\,\mathcal{U}[n]$	$\dfrac{z(z-\cos b)}{z^2-(2\cos b)z+1}$
12	$\sin[bn]\,\mathcal{U}[n]$	$\dfrac{z\sin b}{z^2-(2\cos b)z+1}$
13	$a^n\cos[bn]\,\mathcal{U}[n]$	$\dfrac{z(z-a\cos b)}{z^2-(2a\cos b)z+a^2}$
14	$a^n\sin[bn]\,\mathcal{U}[n]$	$\dfrac{za\sin b}{z^2-(2a\cos b)z+a^2}$

G Linear Algebra

A rectangular table of $m \times n$ numbers is called a matrix:

$$A_{m\times n}=(a_{ij})=\begin{pmatrix} a_{11} & a_{12} & \cdots & a_{1n} \\ a_{21} & a_{22} & \cdots & a_{2n} \\ \vdots & \vdots & & \vdots \\ a_{m1} & a_{m2} & \cdots & a_{mn} \end{pmatrix}$$

where m is the number of rows and n is the number of columns. The numbers a_{ij} ($i=1,2,\ldots,m;\ j=1,2,\ldots,n$) are called matrix elements. The first index i points to the row number and the second index j points to the column number where this element is located.

If the number of rows equals the number of columns ($m=n$), the matrix is called square of the order n.

G.1 Special Matrices

A *diagonal matrix* has zeros everywhere outside of the main diagonal:

$$A = \text{diag}(a_{11}, a_{22}, \ldots, a_{nn}) = \begin{pmatrix} a_{11} & & 0 \\ & a_{22} & \\ & & \ddots \\ 0 & & & a_{nn} \end{pmatrix} = (a_{ij}\delta_{ij})$$

when δ_{ij} is defined by: $\delta_{ij} = \begin{cases} 1, & \text{if } i = j \\ 0, & \text{else} \end{cases}$

An *identity matrix* is a special kind of a diagonal matrix with 1s on the main diagonal ($a_{ii} = 1$) and denoted by I.

An *upper (lower) triangular matrix* is the matrix with zero elements below (above) the main diagonal.

G.2 Matrix Addition and Subtraction

$$\begin{pmatrix} a_{11} & a_{12} & \cdots & a_{1n} \\ a_{21} & a_{22} & \cdots & a_{2n} \\ \vdots & \vdots & & \vdots \\ a_{m1} & a_{m2} & \cdots & a_{mn} \end{pmatrix} \pm \begin{pmatrix} b_{11} & b_{12} & \cdots & b_{1n} \\ b_{21} & b_{22} & \cdots & b_{2n} \\ \vdots & \vdots & & \vdots \\ b_{m1} & b_{m2} & \cdots & b_{mn} \end{pmatrix} = \begin{pmatrix} a_{11} \pm b_{11} & a_{12} \pm b_{12} & \cdots & a_{1n} \pm b_{1n} \\ a_{21} \pm b_{21} & a_{22} \pm b_{22} & \cdots & a_{2n} \pm b_{2n} \\ \vdots & \vdots & & \vdots \\ a_{m1} \pm b_{m1} & a_{m2} \pm b_{m2} & \cdots & a_{mn} \pm b_{mn} \end{pmatrix}$$

G.3 Matrix Determinant

$$\det(A_{2 \times 2}) = \begin{vmatrix} a & b \\ c & d \end{vmatrix} = ad - bc$$

$$\det(A_{3 \times 3}) = \begin{vmatrix} a & b & c \\ d & e & f \\ g & h & i \end{vmatrix} = a\begin{vmatrix} e & f \\ h & i \end{vmatrix} - b\begin{vmatrix} d & f \\ g & i \end{vmatrix} + c\begin{vmatrix} d & e \\ g & h \end{vmatrix}$$

$$\det(A) = |A| = a_{11} \begin{vmatrix} a_{22} & \cdots & a_{2n} \\ \vdots & & \vdots \\ a_{n2} & \cdots & a_{nn} \end{vmatrix} - a_{12} \begin{vmatrix} a_{21} & a_{23} & a_{24} & \cdots & a_{2n} \\ a_{31} & a_{33} & a_{34} & \cdots & a_{3n} \\ \vdots & \vdots & \vdots & & \vdots \\ a_{n1} & a_{n3} & a_{n4} & \cdots & a_{nn} \end{vmatrix} +$$

$$+ a_{13} \begin{vmatrix} a_{21} & a_{22} & a_{24} & \cdots & a_{2n} \\ a_{31} & a_{32} & a_{34} & \cdots & a_{3n} \\ \vdots & \vdots & \vdots & & \vdots \\ a_{n1} & a_{n2} & a_{n4} & \cdots & a_{nn} \end{vmatrix} - \cdots$$

You can reduce the determinant's order from n to $n-1$ using the approach above to simplify the computation.

Determinant properties:

1. Switching two consecutive rows (columns) changes the sign of the determinant.
2. If there are any linearly dependent rows (columns) then the determinant is zero.
3. Adding a linear combination of rows (columns) to some row (column) will not change the determinant.
4. $\det(kA) = k^n \det(A)$, when k is a scalar.
5. For square matrices A and B: $\det(AB) = \det(A)\det(B)$.
6. $\det(I_{m \times m} + A_{m \times n} B_{n \times m}) = \det(I_{n \times n} + B_{n \times m} A_{m \times n})$.
7. $\det(A^{-1}) = \det(A)^{-1}$.

G.4 Singular Matrix

If $\det(A) = 0$, A is called singular (or noninvertible). The matrix is singular if and only if it has linearly dependent rows (or columns).

G.5 Regular Matrix

Nonsingular matrices are called regular matrices.

NOTE The terms "determinant," "regular," and "singular" are defined only for square matrices.

G.6 Transposed Matrix

$A^T = (a_{ji})$ is the transposed matrix $A = (a_{ij})$. Note that in the transposed matrix the rows are switched with the columns.

Properties:

1. $(A^T)^T = A$
2. $(A + B)^T = A^T + B^T$
3. $(kA)^T = kA^T$, k is scalar
4. $(AB)^T = B^T A^T$
5. $\det(A^T) = \det(A)$
6. $\det(sI - A^T) = \det(sI - A)$ (eigenvalues of A and A^T are the same)

G.7 Symmetric Matrix

$$A^T = A$$

G.8 Antisymmetric Matrix

$$A^T = -A$$

G.9 Matrix Multiplication

$$A_{m\times n} \cdot B_{n\times p} = C_{m\times p} = (c_{ij}) = \left(\sum_{k=1}^{n} a_{ik}b_{kj}\right); \quad (i=1,2,\ldots,m; j=1,2,\ldots,p)$$

CAUTION! The matrix multiplication is not always commutative, that is, it could be that $AB \neq BA$. For example, $\begin{pmatrix} 3 & 0 \\ 7 & 0 \end{pmatrix}\begin{pmatrix} 1 & 2 \\ 1 & 2 \end{pmatrix} = \begin{pmatrix} 3 & 6 \\ 7 & 14 \end{pmatrix} \neq \begin{pmatrix} 17 & 0 \\ 17 & 0 \end{pmatrix} = \begin{pmatrix} 1 & 2 \\ 1 & 2 \end{pmatrix}\begin{pmatrix} 3 & 0 \\ 7 & 0 \end{pmatrix}$.

CAUTION! If $AC = BC$, it does not mean that $A = B$. For example, $\begin{pmatrix} 1 & 2 \\ 3 & 4 \end{pmatrix}\begin{pmatrix} 1 & 0 \\ 1 & 0 \end{pmatrix} = \begin{pmatrix} 1 & 2 \\ 4 & 3 \end{pmatrix}\begin{pmatrix} 1 & 0 \\ 1 & 0 \end{pmatrix}$, but $\begin{pmatrix} 1 & 2 \\ 3 & 4 \end{pmatrix} \neq \begin{pmatrix} 1 & 2 \\ 4 & 3 \end{pmatrix}$.

G.10 Powers of the Square Matrix

$$A^k = \underbrace{AA\ldots A}_{k \text{ times}}$$

G.11 Square Root of the Square Matrix (Cholesky Decomposition)

Algorithm for finding the matrix L so that $LL^T = A$ for a positive symmetric matrix A (see definitions below):

$$L_{ii} = \sqrt{A_{ii} - \sum_{k=1}^{i-1} L_{ik}^2} \quad i=1,\ldots,n \qquad L_{ji} = \frac{1}{L_{ii}}\left(A_{ij} - \sum_{k=1}^{i-1} L_{ik}L_{jk}\right) \quad j=i+1,\ldots,n$$

G.12 Matrix Exponent

For a scalar power, the exponent could be developed into the following Taylor series:

$$e^x = 1 + x + \frac{x^2}{2!} + \frac{x^3}{3!} + \frac{x^4}{4!} + \cdots$$

For $x = at$ (a is scalar):

$$e^{at} = 1 + at + \frac{(at)^2}{2!} + \frac{(at)^3}{3!} + \frac{(at)^4}{4!} + \cdots$$

Similarly, we could define for a matrix A:

$$e^{At} = I + At + \frac{(At)^2}{2!} + \frac{(At)^3}{3!} + \frac{(At)^4}{4!} + \cdots$$

NOTE The exponent of $n \times n$ matrix A is $n \times n$ matrix too.

If we take the Laplace transform from both sides for a large enough $|s|$:

$$\mathcal{L}\{e^{At}\} = \frac{I}{s} + \frac{A}{s^2} + \frac{A^2}{s^3} + \frac{A^3}{s^4} + \frac{A^4}{s^5} + \cdots = \frac{1}{s}\left(I + \left(\frac{A}{s}\right) + \left(\frac{A}{s}\right)^2 + \left(\frac{A}{s}\right)^3 + \left(\frac{A}{s}\right)^4 + \cdots\right)$$

On the other hand, for scalar x, we know that the Taylor series of $\frac{1}{1-x} = 1 + x + x^2 + x^3 + \cdots$ and for some matrix C it would be $(I - C)^{-1} = I + C + C^2 + C^3 + \cdots$ if this series is converging. Thus, for $C = -\frac{A}{s}$, we get $\mathcal{L}\{e^{At}\} = \frac{1}{s}\left(I - \frac{A}{s}\right)^{-1} = (sI - A)^{-1}$. Now, if we take the inverse Laplace transform from both sides, we get

$$e^{At} = \mathcal{L}^{-1}\{(sI - A)^{-1}\}$$

NOTE The solution of $\dot{x}(t) = Ax(t)$ matrix differential equation is the vector $x(t) = e^{At}x(0)$.

Properties:

- $p(A)e^{At} = e^{At}p(A)$, for any polynomial p
- $e^{A^T t} = (e^{At})^T$
- $e^{-At} = (e^{At})^{-1}$
- $\frac{d}{dt}e^{At} = Ae^{At}$

CAUTION! It is not always true that $e^A e^B = e^{A+B}$ (true only when $AB = BA$). Here is an example: $e^{\begin{pmatrix} 0 & 1 \\ -1 & 0 \end{pmatrix}} e^{\begin{pmatrix} 0 & 1 \\ 0 & 0 \end{pmatrix}} = \begin{pmatrix} 0.54 & 1.38 \\ -0.84 & -0.3 \end{pmatrix}$ and $e^{\begin{pmatrix} 0 & 1 \\ -1 & 0 \end{pmatrix} + \begin{pmatrix} 0 & 1 \\ 0 & 0 \end{pmatrix}} = \begin{pmatrix} 0.16 & 1.4 \\ -0.7 & 0.16 \end{pmatrix}$.

G.13 Matrix Rank

rank(A) is the number of linearly independent rows (columns).

Rows $r_1, r_2, \ldots, r_n$ are linearly independent if their linear combination $\alpha_1 r_1 + \alpha_2 r_2 + \cdots + \alpha_n r_n$ is not equal to zero for any α_i; $i = 1, \ldots, n$ (except the trivial combination $\alpha_1 = \alpha_2 = \cdots = \alpha_n = 0$).

Properties:

- rank(AB) $\leq \min(\text{rank}(A), \text{rank}(B))$.
- Square matrix A has a full rank if and only if A is regular.

G.14 Inverse of a Matrix

The inverse of a matrix A is denoted by A^{-1} and satisfies

$$A(A^{-1}) = (A^{-1})A = I$$

Properties:

1. $(A^{-1})^{-1} = A$
2. $(A^{-1})^T = (A^T)^{-1}$
3. $(AB)^{-1} = B^{-1}A^{-1}$
4. Lemma of matrix inverse: If A and C are regular and $A + BCD$ exists and is invertible, then: $(A + BCD)^{-1} = A^{-1} - A^{-1}B(DA^{-1}B + C^{-1})^{-1}DA^{-1}$
5. Given $A_{n \times n}$, $B_{n \times m}$, $C_{m \times n}$: $(I - C(sI - A)^{-1}B)^{-1} = I + C(sI - A - BC)^{-1}B$

Calculate inverse matrix:

2×2:
$$A = \begin{pmatrix} a & b \\ c & d \end{pmatrix} \Rightarrow A^{-1} = \frac{1}{ad-bc}\begin{pmatrix} d & -b \\ -c & a \end{pmatrix}$$

3×3:
$$A = \begin{pmatrix} a & b & c \\ d & e & f \\ g & h & i \end{pmatrix} \Rightarrow A^{-1} = \frac{1}{\det(A)} \begin{pmatrix} \begin{vmatrix} e & f \\ h & i \end{vmatrix} & -\begin{vmatrix} b & c \\ h & i \end{vmatrix} & \begin{vmatrix} b & c \\ e & f \end{vmatrix} \\ -\begin{vmatrix} d & f \\ g & i \end{vmatrix} & \begin{vmatrix} a & c \\ g & i \end{vmatrix} & -\begin{vmatrix} a & c \\ d & f \end{vmatrix} \\ \begin{vmatrix} d & e \\ g & h \end{vmatrix} & -\begin{vmatrix} a & b \\ g & h \end{vmatrix} & \begin{vmatrix} a & b \\ d & e \end{vmatrix} \end{pmatrix}$$

CAUTION! In solving matrix equations, it is prohibited to reduce matrices on both sides. To simplify the equation, you need to multiply both sides of the equation by some inverse matrix from the left or from the right, and then use $AA^{-1} = I$.

G.15 Trace of a Matrix

For a square matrix $A_{n \times n} = (a_{ij})$, we define: $tr(A) = \sum_{i=1}^{n} a_{ii}$, the trace of the matrix A.

Properties:
- $tr(A+B) = tr(A) + tr(B)$
- $tr(AB) = tr(BA)$

Example

To compute the trace of the following matrix: $\begin{pmatrix} 1 \\ 2 \\ 0 \\ 4 \\ 0 \end{pmatrix} \cdot (1 \ 2 \ 3 \ 0 \ 5)$, we could compute the 5×5 matrix first and then add all elements on the main diagonal. Instead, it is much more convenient to use the property $tr(AB) = tr(BA)$ and get an element-wise multiplication of those two vectors: $tr\left(\begin{pmatrix} 1 \\ 2 \\ 0 \\ 4 \\ 0 \end{pmatrix}(1 \ 2 \ 3 \ 0 \ 5)\right) = tr\left((1 \ 2 \ 3 \ 0 \ 5)\begin{pmatrix} 1 \\ 2 \\ 0 \\ 4 \\ 0 \end{pmatrix}\right) =$ $1 \cdot 1 + 2 \cdot 2 + 0 \cdot 3 + 4 \cdot 0 + 0 \cdot 5 = 5.$

G.16 Orthogonal Matrix
$A^T = A^{-1}$ (A is a real matrix).

G.17 Orthonormal Matrix
The orthonormal matrix is an orthogonal matrix with the columns normalized to the norm 1.

G.18 Eigenvalues and Eigenvectors of a Square Matrix
For a matrix $A_{n \times n}$, we define the *eigenvalues* as the scalar solutions s_i of the equation $Ax_i = s_i x_i$ for some vector x_i. The vector x_i that belongs to the eigenvalue s_i is called *eigenvector*.

Computation: The polynomial $\det(sI - A)$ is called the *characteristic polynomial* $\Delta(s)$ of A, and its roots are the eigenvalues of A.

By solving the equation $\det(sI - A) = 0$ we find all the eigenvalues $\{s_i\}$, and then by solving the system of equations $Ax_i = s_i x_i$ we find all the eigenvectors $\{x_i\}$.

Diagonalization matrix: Construct a diagonalization transform matrix T from the eigenvectors as its columns. You may apply this similarity transform to the matrix A to get: $T^{-1}AT = \mathrm{diag}\{s_i\}$.

Properties:

1. $tr(A) = \sum_{i=1}^{n} s_i$.
2. $\det(A) = \prod_{i=1}^{n} s_i$.
3. $A - s_i I$ is singular.
4. For 2×2 matrices: $\Delta(s) = \det(sI - A) = s^2 - (tr(A))s + \det(A)$.
5. Any real symmetric matrix A is diagonizable, all its eigenvalues are real, and all its eigenvectors are orthogonal.
6. $s = 0$ is an eigenvalue of A if and only if A is singular (noninvertible).
7. $s \neq 0$ is an eigenvalue of A if and only if subtracting s from the main diagonal will reduce the rank(A).
8. All eigenvalues of upper or lower triangular matrices are located on the main diagonal.
9. If $f(A)$ is an analytic function of A that can be written as polynomial series (most functions are …), then for any invertible transform T, $T^{-1}f(A)T = f(T^{-1}AT)$. This is an effective technique to compute complicated matrix functions.
10. A and A^T have the same eigenvalues.

Example

Let's say we want to compute $\begin{pmatrix} 1 & 2 \\ 0 & 2 \end{pmatrix}^{2000}$ without using a computer. To solve that problem, we will define $A = \begin{pmatrix} 1 & 2 \\ 0 & 2 \end{pmatrix}$ and compute its eigenvalues s_i and eigenvectors x_i.

This will allow us to find a diagonalizing similarity transform T such that $T^{-1}AT$ is diagonal. Then, we will use the identity $A^{2000} = (T\,\text{diag}(s_i)T^{-1})^{2000}$ to compute the matrix.

To compute eigenvalues, $\det(sI - A) = \begin{vmatrix} s-1 & -2 \\ 0 & s-2 \end{vmatrix} = (s-1)(s-2) = 0$, thus $s_1 = 1$ and $s_2 = 2$. We could get those values using property 8 of eigenvalues without any computations. Now, we need to compute the eigenvectors. For that, we solve the system of equations $Ax_i = s_i x_i$ for each s_i:

$$\begin{pmatrix} 1 & 2 \\ 0 & 2 \end{pmatrix} \begin{pmatrix} x_1 \\ x_2 \end{pmatrix} = s_1 \begin{pmatrix} x_1 \\ x_2 \end{pmatrix} = \begin{pmatrix} x_1 \\ x_2 \end{pmatrix} \Rightarrow \begin{cases} x_1 + 2x_2 = x_1 \\ 2x_2 = x_2 \end{cases} \Rightarrow x_2 = 0$$

Note that we should always get an infinite number of solutions if the matrix is diagonalizable. In that case, we can choose any x_1, for example, $x_1 = 1$, and get $x_1 = \begin{pmatrix} 1 \\ 0 \end{pmatrix}$.

Similarly,

$$\begin{pmatrix} 1 & 2 \\ 0 & 2 \end{pmatrix} \begin{pmatrix} x_1 \\ x_2 \end{pmatrix} = s_2 \begin{pmatrix} x_1 \\ x_2 \end{pmatrix} = 2 \begin{pmatrix} x_1 \\ x_2 \end{pmatrix} \Rightarrow \begin{cases} x_1 + 2x_2 = 2x_1 \\ 2x_2 = 2x_2 \end{cases} \Rightarrow x_1 = 2x_2$$

Again, we have one degree of freedom and we can choose any x_2. For example, $x_2 = \begin{pmatrix} 2 \\ 1 \end{pmatrix}$.

The diagonalizing transform is $T = (x_1 | x_2) = \begin{pmatrix} 1 & 2 \\ 0 & 1 \end{pmatrix}$, thus $T^{-1} = \begin{pmatrix} 1 & -2 \\ 0 & 1 \end{pmatrix}$, and the diagonal matrix of eigenvalues is $D = \text{diag}(s_i) = \begin{pmatrix} 1 & 0 \\ 0 & 2 \end{pmatrix}$.

Finally, $A^{2000} = (TDT^{-1})^{2000} = \underbrace{(TDT^{-1})(TDT^{-1})(TDT^{-1}) \cdots (TDT^{-1})}_{2000 \text{ times}} = TD^{2000}T^{-1} =$

$\begin{pmatrix} 1 & 2 \\ 0 & 1 \end{pmatrix} \begin{pmatrix} 1 & 0 \\ 0 & 2 \end{pmatrix}^{2000} \begin{pmatrix} 1 & -2 \\ 0 & 1 \end{pmatrix} = \begin{pmatrix} 1 & 2 \\ 0 & 1 \end{pmatrix} \begin{pmatrix} 1^{2000} & 0 \\ 0 & 2^{2000} \end{pmatrix} \begin{pmatrix} 1 & -2 \\ 0 & 1 \end{pmatrix} = \begin{pmatrix} 1 & 2^{2001} - 2 \\ 0 & 2^{2000} \end{pmatrix}$.

Cayley-Hamilton Theorem

If $\det(sI - A) = s^n + \alpha_1 s^{n-1} + \cdots + \alpha_n$, then $A^n + \alpha_1 A^{n-1} + \cdots + \alpha_n I = 0$. In other words, matrix A is a zero of its own characteristic polynomial.

The polynomial of the lowest degree $p(s)$ such that $p(A) = 0$ is called a *minimal polynomial* of A.

G.19 Similar Matrices

Matrices A and B are *similar*, if there exists an invertible transform T such that $A = T^{-1}BT$ (alternatively, $TA = BT$).

Properties:

- Similar matrices have the same eigenvalues.
- Matrix A is diagonalizable if and only if there exists the invertible transformation T such that $T^{-1}AT$ is diagonal, and the columns of T are the eigenvectors of A.
- A is similar to A^T.

Example

We will find a similarity transform between A and A^T if they are diagonalizable. Let's assume that T_1 is a diagonalization transform of A and T_2 is a diagonalization transform of A^T. Since both A and A^T have the same eigenvalues diagonal matrix D, $T_1^{-1}AT_1 = D = T_2^{-1}A^T T_2$. Now, we multiply both sides of the equation by T_1 on the left and T_1^{-1} on the right: $T_1 T_1^{-1} A T_1 T_1^{-1} = A = T_1 T_2^{-1} A^T T_2 T_1^{-1}$. So, if we define the similarity transform $P = T_2 T_1^{-1}$, then $A = P^{-1} A^T P$.

G.20 Positive Definite and Semidefinite Matrices

A real symmetric matrix $A_{n \times n}$ is called *positive definite* if for any vector $x \neq 0$ the quadratic form is positive, that is, $x^T A x = \sum_{i=1}^{n}\sum_{j=1}^{n} a_{ij} x_i x_j > 0$. We write that as $A \succ 0$.

A real symmetric matrix $A_{n \times n}$ is called *positive semidefinite* if for any vector $x \neq 0$ the quadratic form is positive, that is, $x^T A x = \sum_{i=1}^{n}\sum_{j=1}^{n} a_{ij} x_i x_j \geq 0$.

Properties:

1. *Sylvester's Criterion:* If for the symmetric matrix A, all the main minors are positive $a_{11} > 0$, $\begin{vmatrix} a_{11} & a_{12} \\ a_{12} & a_{22} \end{vmatrix} > 0$, $\begin{vmatrix} a_{11} & a_{12} & a_{13} \\ a_{12} & a_{22} & a_{23} \\ a_{13} & a_{23} & a_{33} \end{vmatrix} > 0 \ldots$, then A is positive definite.

2. If all the main minors are nonnegative, then symmetric A is positive semidefinite.

3. A square matrix is symmetric and positive definite if and only if there exists invertible matrix D such that $A = D^T D$.

4. A square matrix of the order n is symmetric and positive semidefinite if and only if there exists matrix D (rank$(D) < n$) such that $A = D^T D$.

5. Matrix A is positive definite (semidefinite) if and only if all its eigenvalues are positive (nonnegative).

G.21 Block Matrices

1. $\begin{pmatrix} A_{11} & A_{12} \\ A_{21} & A_{22} \end{pmatrix} \begin{pmatrix} B_{11} & B_{12} \\ B_{21} & B_{22} \end{pmatrix} = \begin{pmatrix} A_{11}B_{11} + A_{12}B_{21} & A_{11}B_{12} + A_{12}B_{22} \\ A_{21}B_{11} + A_{22}B_{21} & A_{21}B_{12} + A_{22}B_{22} \end{pmatrix}$

2. For square matrices A and D: $\det \begin{pmatrix} A & 0 \\ C & D \end{pmatrix} = \det \begin{pmatrix} A & B \\ 0 & D \end{pmatrix} = \det(A)\det(D)$.

NOTE The eigenvalues of a block-triangular matrix are the eigenvalues of the main diagonal blocks.

3. If all matrices in the blocks exist,

$$\begin{pmatrix} A & B \\ C & D \end{pmatrix}^{-1} = \begin{pmatrix} A^{-1} - A^{-1}B(CA^{-1}B - D)^{-1}CA^{-1} & A^{-1}B(CA^{-1}B - D)^{-1} \\ (CA^{-1}B - D)^{-1}CA^{-1} & -(CA^{-1}B - D)^{-1} \end{pmatrix} =$$

$$= \begin{pmatrix} -(BD^{-1}C - A)^{-1} & (BD^{-1}C - A)^{-1}BD^{-1} \\ D^{-1}C(BD^{-1}C - A)^{-1} & D^{-1} - D^{-1}C(BD^{-1}C - A)^{-1}BD^{-1} \end{pmatrix}$$

4. If A is invertible, then $\det\begin{pmatrix} A & B \\ C & D \end{pmatrix} = \det(A)\det(D - CA^{-1}B)$.

5. If A and D are invertible, then

$$\begin{pmatrix} A & 0 \\ C & D \end{pmatrix}^{-1} = \begin{pmatrix} A^{-1} & 0 \\ -D^{-1}CA^{-1} & D^{-1} \end{pmatrix}; \quad \begin{pmatrix} A & B \\ 0 & D \end{pmatrix}^{-1} = \begin{pmatrix} A^{-1} & -A^{-1}BD^{-1} \\ 0 & D^{-1} \end{pmatrix}$$

Example
For the following matrix $M = \begin{pmatrix} 9 & -1 & 5 \\ 8 & 3 & 2 \\ 0 & 0 & 3 \end{pmatrix}$, we need to compute $\det(M)$, M^{-1}, and prove that $M \succ 0$.

We divide the matrix into four blocks as follows:

$$M = \left(\begin{array}{cc|c} 9 & -1 & 5 \\ 8 & 3 & 2 \\ \hline 0 & 0 & 3 \end{array} \right)$$

The blocks are $A = \begin{pmatrix} 9 & -1 \\ 8 & 3 \end{pmatrix}$, $B = \begin{pmatrix} 5 \\ 2 \end{pmatrix}$, $C = (0 \ \ 0)$, $D = (3)$. Since C is a block of zeros, the matrix is block triangular. Therefore,

$$\det(M) = \det(A)\det(D) = (9 \cdot 3 + 8) \cdot 3 = 105$$

$$M^{-1} = \begin{pmatrix} A^{-1} & -A^{-1}BD^{-1} \\ 0 & D^{-1} \end{pmatrix} = \left(\begin{array}{c|c} \frac{1}{35}\begin{pmatrix} 3 & 1 \\ -8 & 9 \end{pmatrix} & -A^{-1}BD^{-1} \\ \hline 0 & 1/3 \end{array} \right) = \begin{pmatrix} 3/35 & 1/35 & -0.162 \\ -8/35 & 9/35 & 0.21 \\ 0 & 0 & 1/3 \end{pmatrix}$$

It would be easy to use the Sylvester criterion to check if the matrix is positive definite, but we cannot do it because the matrix is not symmetrical. We need to find the eigenvalues (which are the eigenvalues of the main diagonal blocks for block triangular matrix). These eigenvalues are 3, 5, and 7. All eigenvalues are positive, thus $M \succ 0$.

H Random Variables

Probability is a number between 0 and 1 that shows the odds of some event, and where numbers closer to 1 mean stronger likelihood. In simple cases, probability could be approximated by estimating the ratio between the number of desired outcomes out of all possible outcomes.

H.1 Expected Value (Mean)

Given a random signal X that could have possible values (outcomes) $x_1, x_2, \ldots, x_n, \ldots$ and given probability of each one of those values $p_1, p_2, \ldots, p_n, \ldots$ described by a given distribution, the *expected value* (or mean, or average) for discrete random variables is given by

$$E[X] = \sum_i p_i x_i = p_1 x_1 + p_2 x_2 + \cdots + p_n x_n + \cdots$$

which is the weighted sum of x_i values, where $\sum_i p_i = 1$.

For example, if there are 10 possible values that X can take, and all of them are equally likely, then the expected value will be $E[X] = \sum_i \frac{1}{10} x_i = \frac{1}{10} \sum_i x_i$, which is an arithmetic mean.

For continuous random variables, and a given density function $p(x)$,

$$E[X] = \int_{-\infty}^{\infty} x p(x) dx$$

Properties:

- If $X = Y$, then $E[X] = E[Y]$
- $E[\alpha X + \beta Y] = \alpha E[X] + \beta E[Y]$
- $E\left[\sum_{i=1}^{\infty} X_i\right] = \sum_{i=1}^{\infty} E[X_i]$

H.2 Variance

A *variance* is the amount of deviation from the mean μ. It is defined by

$$\text{Var}(X) = E[(X - \mu)^2]$$

Because of the linearity property of the expected value, it could be shown that

$$\text{Var}(X) = E[X^2] - (E[X])^2$$

A *standard deviation* is a square root of variance:

$$\sigma_X = \sqrt{\text{Var}(X)}$$

H.3 Gaussian Distribution

There are multiple possible distributions available to describe random processes and noises. The most popular is *Gaussian (normal) distribution*, which is given by (in 1D):

$$f_X(x) = \frac{1}{\sqrt{2\pi\sigma^2}} e^{-\frac{(x-\mu)^2}{2\sigma^2}}$$

where μ is distribution's mean and σ is distribution's standard deviation. It is denoted by $x \sim N(\mu, \sigma)$.

Multivariate distribution for a vector x is given by

$$f_X(x) = \frac{1}{\sqrt{(2\pi)^n |\Sigma|}} \exp\left(-\frac{1}{2}(x-\mu)^T \Sigma^{-1}(x-\mu)\right)$$

where $x = (x_1, x_2, \ldots, x_n)^T$, $\mu = (E[x_1], E[x_2], \ldots, E[x_n])^T$, $\exp(y) = e^y$, Σ is the covariance matrix, and $|\Sigma| = \det(\Sigma)$.

The covariance matrix elements are defined by $(\Sigma_{i,j}) = E[(x_i - E[x_i])(x_j - E[x_j])]$:

$$\Sigma = \mathrm{Cov}(x)$$

$$= \begin{pmatrix} E[(x_1 - E[x_1])^2] & E[(x_1 - E[x_1])(x_2 - E[x_2])] & \cdots & E[(x_1 - E[x_1])(x_n - E[x_n])] \\ E[(x_2 - E[x_2])(x_1 - E[x_1])] & E[(x_2 - E[x_2])^2] & \cdots & E[(x_2 - E[x_2])(x_n - E[x_n])] \\ \vdots & \vdots & \ddots & \vdots \\ E[(x_n - E[x_n])(x_1 - E[x_1])] & E[(x_n - E[x_n])(x_2 - E[x_2])] & \cdots & E[(x_n - E[x_n])^2] \end{pmatrix}$$

References

Ackermann, J. (1972). "Der Entwurf linearer Regelungssysteme im Zustandsraum." *Regelungstechnik,* 20: 291–300.

Bass, R. W., and Gura, I. (1965). "High Order Design via State-Space Considerations." Proceedings of the 1965 Joint Automatic Control Conference, Troy, NY, pp. 311–318.

Baum, L. E., and Petrie, T. (1966). "Statistical Inference for Probabilistic Functions of Finite State Markov Chains." *The Annals of Mathematical Statistics,* 37(6): 1554–1563.

Hurwitz, A. (1895). "Ueber die Bedingungen, unter welchen eine Gleichung nur Wurzeln mit negativen reellen Theilen besitzt." *Mathematische Annalen,* 46(2): 273–284.

Kalman, R. E. (1960). "A New Approach to Linear Filtering and Prediction Problems." *Journal of Basic Engineering,* 82: 35.

Kalman, R. E., and Bucy, R. (1961). "New Results in Linear Filtering and Prediction Theory." *ASME Journal of Basic Engineering,* 83: 95–108.

Letov, A. M. (1960). "Analytical Design of Controllers" (in Russian). *Automation and Remote Control,* 4: 436–441; 5: 561–568; 6: 661–665.

Luenberger, D. G. (1964). "Observing the State of a Linear System." *IEEE Transactions on Military Electronics,* MIL-8: 74–80.

Lyapunov, A. M. (1892). *The General Problem of the Stability of Motion* (in Russian). Doctoral dissertation, Kharkov University.

Routh, E. J. (1877). *A Treatise on the Stability of a Given State of Motion, Particularly Steady Motion.* London: Macmillan and Co.

Stratonovich, R. L. (1959). "On the Theory of Optimal Non-Linear Filtering of Random Functions." *Theory of Probability and Its Applications*, 4: 223–225.

Index

Note: Page numbers followed by *f* denote figures.

A

Ackermann formula, 37, 46
Actuators, plant structure with, 2, 3*f*
Algebra. *See also* Linear algebra
    inequalities, 148–149
    polynomials, 149–150
Algebraic Riccati equation (ARE):
    continuous-time (CARE), 88–89, 122
    discrete-time (DARE), 89, 109
Amplitude spectrum, 153
Antisymmetric matrix, 162
ARE. *See* Algebraic Riccati equation
Asymptotic stability, 7–9, 79, 81–82
    Lyapunov function for, 80–81
    Routh-Hurwitz criterion for, 9

B

Basis, completing to a, 58–59
Bass-Gura formula, 37, 42, 46
BIBO stability. *See* Bounded-input bounded-output stability
Block matrices, 168–169
Bode diagram, 3
Bounded-input bounded-output (BIBO) stability, 7–8
Bridge crane control:
    linearization for, 142
    model for, 140–142, 140*f*
    system description for, 140–141, 140*f*
Bucy, Richard, 105

C

Calculus:
    definite integrals, 151
    derivative rules, 150
    derivative table, 150
    finite series, 152

Calculus (*Cont.*):
    indefinite integrals, 151
    infinite series, 152
    integration rules, 151
    Taylor series, 152
Canonical forms:
    controller, 27–28
    diagonal, 29
    Jordan, 29
    noncontrollable, 57–58, 64
    nonobservable, 58–59
    observer, 28–29
    state-space representations, 27–29, 57–59
CARE. *See* Algebraic Riccati equation
Cauchy inequality, 149
Causal system, 20, 27
Cayley-Hamilton theorem, 167
Characteristic polynomial, 19, 21, 26, 37, 42, 46, 96, 166
    in closed loop, 37, 46, 96
    eigenvalues of matrix computed from, 19, 21
    in open loop, 19
    transfer function denominator as, 19
Cholesky decomposition. *See* Square root of square matrix
Closed-loop system, 2*f*
    characteristic polynomial in, 19, 37, 46, 96
    eigenvalues, 19, 21
    state equations, 36
    state feedback control architecture, 36*f*, 46*f*, 47*f*
    transfer function of, 10, 36, 38, 48–49
Complex Fourier coefficients, 153
Complex Fourier series, 153
Complex numbers:
    Cartesian representation of, 148
    de Moivre's formula for, 148
    division of, 148

Complex numbers (*Cont.*):
  Euler's theorem for, 148
  multiplication of, 148
  polar representation of, 148
Continuous-time algebraic Riccati equation.
    See Algebraic Riccati equation
Control effort, 1, 2*f*, 122–123
Control system:
  design goals for, 1–2, 2*f*
  modeling, 3–4, 4*f*
Controllability:
  of eigenvalues, 59–60
  Gramian of, 85
  matrix, 24
  of state-space representations, 24–25
Controller canonical form. *See* Canonical forms
Convolution integral, 154
Convolution sum, 154
Cost function:
  cross-product extension of, 88–89
  optimal for LQR, 88–89
  optimal for stationary LQG, 123
  prescribed degree of stability extension, 89
  standard, 87
Covariance matrix, 110, 115, 171

D

DARE. *See* Algebraic Riccati equation
DC motor model, 3–4
de Moivre's formula, 148
Definite integrals, 151
Derivative rules, 150
Derivative table, 150
Detectability, 59–60
Determinant, 161–162
DFT. *See* Discrete Fourier transform
Diagonal canonical form, 29
Diagonal matrix, 161
Diagonalization, 166
Differential equations, 4–5, 13, 20, 158–159
Dirac delta function, 154
Direct Lyapunov method, 80–81
Discrete algebraic Riccati equation (DARE), 89, 109
Discrete Fourier transform (DFT), 154
Double inverted pendulum:
  linearization for, 136
  model for, 134–136
  system description for, 133–134, 133*f*
Duality principle, 121–122

E

Eigenvalues, 21–22, 26, 166–167
Eigenvectors, 166–167

Equilibrium points:
  defined, 69
  linearization with, 69–71
  Lyapunov stable, 79, 81, 85
  not in origin, 81, 83
Euler theorem, 148
Expected value, 170

F

Final value theorem, 157, 159
Finite series, 152
Fourier series, 153
Fourier transform (FT):
  common pairs, 156
  definition, 155
  discrete, 154
  properties, 155

G

Gaussian distribution, 109, 170–171
Globally asymptotically stable systems, 81
Gramian of controllability, 85

H

Hölder's inequality, 149
Homogeneous linear differential equation, 159

I

Identity matrix, 161
Indefinite integrals, 151
Inequalities:
  Cauchy, 149
  Hölder's, 149
  triangle, 148–149
Infinite horizon LQG, 122
Infinite series, 152
Integration, 151
Integrator in the loop, 38–39, 39*f*
Internally stable systems, 79–80, 80*f*
Inverse of matrix, 164–165
Inverted pendulum on cart, 75–77, 75*f*
Invertible matrix, 164–165

J

Jacobian matrix, 71
Jordan canonical form, 29

K

Kalman, Rudolf, 105
Kalman-Bucy filter, 105, 121–123, 125

Kalman filter:
    alternative formulation for, 111–112, 112f
    equations, 105–108, 106f, 107f, 108f, 109f
    measurement update with, 110–111
    stationary, 111
    steady-state, 111
    time update with, 110–111
Kronecker delta function, 154

L

Laplace transform (LT):
    common pairs, 158
    definition, 157
    inverse, 6–8, 18, 20, 157
    properties, 157
Letov theorem, 95–97
Linear algebra, 160–169
    antisymmetric matrix, 162
    block matrices, 168–169
    diagonal matrix, 161
    eigenvalues, 166–167
    eigenvectors of square matrix, 166–167
    identity matrix, 161
    inverse of matrix, 164–165
    matrix addition and subtraction, 161
    matrix determinant, 161–162
    matrix exponent, 163–164
    matrix multiplication, 163
    matrix rank, 164
    matrix trace, 165
    matrix transpose, 162
    orthogonal matrix, 166
    orthonormal matrix, 166
    positive definite matrix, 168
    positive semidefinite matrix, 168
    powers of square matrix, 163
    regular matrix, 162
    similar matrix, 167–168
    singular matrix, 162
    square root of square matrix, 163
    symmetric matrix, 162
    triangular matrix, 161
Linear combination, 24, 162, 164
Linear independence, 164
Linear quadratic Gaussian (LQG) control:
    block diagram for, 122f, 123f
    description, 122–123, 122f, 123f
    infinite horizon, 122
    Kalman-Bucy filter for, 121–123, 125
    optimal cost function for stationary, 123
    separation principle with, 122–123
    state equations of, 123
Linear quadratic regulator (LQR):
    continuous-time, 88
    cost function for, 87–88

Linear quadratic regulator (LQR) (*Cont.*):
    cross-product extension of cost function with, 88–89
    discrete-time, 89
    prescribed degree of stability extension of cost function with, 89
Linear systems, 153
Linear time invariant (LTI) systems, 153
Linearization, 69–71
Lower triangular matrix, 161
LQG control. *See* Linear quadratic Gaussian control
LQR. *See* Linear quadratic regulator
LT. *See* Laplace transform
LTI. *See* Linear time invariant systems
Luenberger, David, 44, 105
Lyapunov, Alexandr, 79
Lyapunov stability:
    continuous-time LTI systems with, 81–82
    direct method for, 80–81
    discrete-time LTI systems with, 82–83
    function, 80–81
    internally stable, 79–80, 80f
    matrix equation, 82–83, 85, 93
    second method for, 80–81
    theorem, 71, 72, 74, 81–82
    2D state space with, 80f

M

Magnetic levitation system control:
    linearization for, 132
    model for, 130–132, 130f
    system description for, 129–130
Matrix. *See* Linear algebra
Measurement noise, 108
Measurement update, 110–111
MIMO systems. *See* Multiple-input multiple-output systems
Minimal realization, 25, 29
Modeling:
    bridge crane control, 140–142, 140f
    control system, 3–4, 4f
    direct current motor, 3–4, 4f
    double inverted pendulum, 133–136
    inverted pendulum, 75
    magnetic levitation system model, 129–132, 130f
    Van der Pol oscillator, 71
Multiple-input multiple-output (MIMO) systems, 21, 88
    controllability of, 25
    controller design with, 11, 38
    stabilizability of, 60
Multivariable Taylor series, 152

N

Nonlinear differential equations, 69
Nonlinear systems:
    differential equations describe, 69, 135
    equilibrium points for, 69
    Lyapunov stability for, 79, 81
Notation and nomenclature, 147

O

Observability:
    definition, 25
    matrix, 25
    matrix of nonobservable canonical form, 59
    matrix of observer canonical form, 29
    of similar system, 26
Observability subspace, 25
Observer, 43–49, 44f, 46f, 47f
Observer canonical forms, 28–29
ODEs. *See* Ordinary differential equations
Optimal controller:
    continuous-time, 88, 93
    cost function minimized for, 91
    discrete-time, 89
    sufficient and necessary conditions for uniqueness of, 88
Optimal observer in presence of noise. *See* Kalman filter
Order of system, 11, 19
Ordinary differential equations (ODEs), 158–159
    conversion from transfer function to, 6–7
    conversion to state-space from, 11–13
    conversion to transfer function from, 4–6
Orthogonal matrix, 166
Orthonormal matrix, 166
Overshoot, 9–10, 10f

P

Parseval's theorem, 153
Partial fractions expansion, 156
Performance specifications in complex domain, 9–10, 10f
Phase spectrum, 153
Plant, 1–2, 2f, 3f
Pole placement:
    with integrator in loop, 38–39, 39f
    state-space controller design by, 35–37, 36f
    tracking input signal using, 38
Polynomials. *See also* Characteristic polynomial
    factorization theorem, 149
    Vieta's formulas, 150
Positive definite matrices, 168
Positive semidefinite matrices, 168
Power spectrum, 153
Powers of square matrix, 163
Proper system, 27

Q

Quadratic form, 88, 168

R

Random variables:
    expected value, 170
    Gaussian distribution, 170–171
    variance, 170
Rank, 164
Regular matrix, 162
Riccati equations, 88–89, 93, 95, 109, 122, 125
Root locus (RL), 96. *See also* Symmetric root locus
Routh-Hurwitz criterion, 9, 16, 72

S

Second method of Lyapunov, 80–81
Separation principle, 47, 121
    LQG control with, 122–123, 123f
    state estimation with, 47–48
Settling time, 9–10, 10f
Signal energy and power, 155
Similarity transform, 26, 42, 48, 166–168
    canonical noncontrollable form with, 57–58
    canonical nonobservable form with, 58–59
    state-space representations, 25–28
Singular matrix, 162
Square root of square matrix (Cholesky decomposition), 163
SRL. *See* Symmetric root locus
Stability:
    asymptotic, 7–9, 79, 81–82
    BIBO, 7–8, 79
    continuous-time systems, 21
    discrete-time systems, 21–22
    internal, 79–80, 80f
    linear quadratic regulator with prescribed degree of, 89
    Lyapunov, 79–85, 80f
    MIMO systems, 60
    Routh-Hurwitz, 9
Stabilizability, 57, 59
Standard deviation, 170
State:
    difference equation, 20
    differential equation, 17–18
    equation solution, 20
    equilibrium, 69
    estimation, 43–49, 44f, 46f, 47f
    feedback control, 35–37, 36f
    variable, 11
    vector, 17
State-space representations:
    block diagrams of, 22–23, 23f
    canonical forms for, 27–30
    of closed-loop feedback control architecture, 36f

Index

State-space representations (*Cont.*):
  control system in, 10–13
  controllability of, 24–25
  conversion from ODE, 11–13
  minimal, 25
  observability of, 25
  similarity transform for, 25–27
  stability, 21
Stationary (steady-state) Kalman filter, 111
Steady-state gain, 38, 111
Stratonovich, Ruslan, 105
Stratonovich filter, 105
Sylvester's criterion, 168
Symmetric matrix, 162
Symmetric root locus (SRL):
  continuous-time, 95–98
  discrete-time, 97–98
System's noise, 108

T

Taylor series, 152
Time invariant, 153
Time update, 110, 111
Trace of matrix, 165
Transfer function:
  characteristic polynomial as denominator of, 19
  closed-loop, 10, 36, 48–49
  of continuous-time system, 19
  conversion from ODE to, 4–6

Transfer function (*Cont.*):
  conversion to ODE from, 6–7
  of discrete-time system, 20
  matrix, 21
  open-loop, 99
  of SISO, 18–20
Transposed matrix, 162
Triangle inequality, 148–149
Triangular matrix, 161
Trigonometric Fourier series, 153
Trigonometric identities, 147

U

Upper triangular matrix, 161

V

Van der Pol oscillator equation, 71
Variance, 170
Vieta's formulas, 150

Z

Z-transform:
  common pairs, 160
  definition, 159
  properties, 159
Zero input response (ZIR), 18
Zero state response (ZSR), 4–6, 8, 18
ZIR. *See* Zero input response
ZSR. *See* Zero state response